Development of Predictive Tools for Membrane Ageing

Pierre LeClech

UNESCO Centre for Membrane Science and Technology, University of New South Wales

Final Report – WaterRA Project 2008

DISCLAIMER

Water Research Australia Limited, is a not for profit company funded by the Australian water industry. (The name of Water Quality Research Australia (WQRA) was changed to Water Research Australia in July 2013)

Water Research Australia and individual contributors are not responsible for the outcomes of any actions taken on the basis of information in this research report, nor for any errors and omissions.

Water Research Australia and individual contributors disclaim all and any liability to any person in respect of anything, and the consequences of anything, done or omitted to be done by a person in reliance upon the whole or any part of this research report.

This research report does not purport to be a comprehensive statement and analysis of its subject matter, and if further expert advice is required, the services of a competent professional should be sought.

Location:

WaterRA Head Office
Level 3, 250 Victoria Square, Adelaide SA 5000

Postal Address:

GPO BOX 1751, Adelaide SA 5001

For more information about WQRA visit the website www.waterra.com.au

Final Report – Development of Predictive Tools for Membrane Ageing

Final Report 2008

ISBN 978-1-921732-12-6

Co-published by IWA Publishing
Alliance House, 12 Caxton Street, London SW1H 0QS, UK
Tel. +44 (0)20 7654 5500, Fax +44 (0)20 7654 5555
publications@iwap.co.uk
iwapublishing.com

ISBN: 978178046251 (Paperback)
ISBN: 978178046558 (eBook)

FOREWORD

Research Report Title: Development of Predictive Tools for Membrane Ageing

Project Leader: Pierre Le Clech

Collaborators: Dr Ho Kyong Shon

Research Nodes: UNSW, UTS

WaterRA Project No. 2008 – Development of Predictive Tools for Membrane Ageing

EXECUTIVE SUMMARY

This study was designed to increase our current understanding on the degradation/ageing mechanisms occurring on porous membranes used in the water and wastewater industries. Accelerated membrane degradation was obtained through both static and consecutive ageing protocols on the pilot-scale, and a range of carefully selected characterisation and analytical techniques was used to characterise the nascent changes faced by the membrane material.

This project was separated into four interrelated sections: (1) critical assessment of characterisation techniques, (2) static accelerated ageing, (3) consecutive accelerated ageing and (4) consecutive ageing of industrially-aged membranes. **The full reports on these sections of the study are included as Appendices 1-4.**

This final report summarises the aims, objectives, outcomes and limitations of the individual work packages, along with some recommendations for future work.

CONTENTS

FIGURES

TABLES

ABBREVIATIONS

FEEM	- fluorescence emission-excitation matrix
FTIR	- fourier transform infrared spectroscopy
HF	- hollow fibre
HPI	- hydrophilic
HPO	- hydrophobic
LC-OCD	- liquid chromatography-organic carbon detection
MBR	- membrane bioreactors
MF	- microfiltration
NaOCl	- sodium hypochlorite
PA	- polyamide
PP	- polypropylene
PS/PES	- polystyrene/polyethersulfone
PVDF	- polyvinylidene fluoride
PVP	- polyvinylpyrrolidone
R_m	- membrane resistance
SDS	- cleaning with surfactants
TGA	- thermogravimetric analysis
TMP	- transmembrane pressure
TOC	- total organic carbon
UF	- ultrafiltration

1. INTRODUCTION

The demand for microfiltration (MF) and ultrafiltration (UF) in the water and wastewater treatment industry has increased greatly since the mid 1990s, due to the requirement for technology to comply with new disinfection standards for drinking water and more stringent regulations for wastewater discharge, respectively [1]. UF membranes, for example, have been reported to combine high pathogen removal ability with minimal energy consumption [2]. Additionally, application of MF/UF membranes in membrane bioreactors (MBR) has sharply increased in the recent years due to their ability to combine biological treatment and filtration in one compact process. A survey on MBR global market reports that this technology is expected to increase from $216.6 million in 2005 to $488 million in 2013 [3]. MF/UF has also been considered as a promising alternative for the pre-treatment to seawater desalination in the last ten years. MF/UF pre-treatment presents various advantages over the conventional pre-treatment techniques, such as environmental friendship in terms of limiting the sludge amount and use of chemicals, lower foot print and cost effectiveness.

Membrane fouling is the term used to describe the undesirable formation of deposits on membrane surfaces. Various membrane fouling types can be broadly classified as organic fouling, inorganic fouling, particulate deposition (colloidal) and biofouling [4]. Some of the important mechanisms of membrane fouling are, adsorption of feed components, clogging of pores, chemical interaction between solutes and membrane material, gel formation, and bacterial growth [5]. In general, fouling can be controlled by suitable pre-treatment of the feed and membrane cleaning strategies. Over long-term operation, membrane fouling is not completely reversible by simple physical cleaning (i.e. backwash and relaxation) therefore, chemical cleaners are used to remove remaining foulants from the membrane surface and restore the permeate flux. On the other hand, cleaning chemicals can potentially lead to removal or degradation of the membrane material, additives and alteration of membrane properties [6]. Generally, polymeric membranes are naturally hydrophobic and must be blended with additives and pore-formers to increase their hydrophilicity such as polyvinylpyrrolidone (PVP). However, the major concern associated with chemical cleaning remains the potential membrane degradation (or ageing) upon repeated fouling-cleaning cycles.

Although previous reports have reported possible chemical reactions between chemical agents and membrane, little is known on the nascent changes occurring at the molecular and structural level of membrane in progressive membrane ageing over prolonged period. Therefore, this study aimed to:

(1) adapt existing characterisation techniques capable of detecting nascent changes of membrane material to determine the loss of membrane integrity,

(2) carry out an arbitrary ranking of membrane stability against different chemical cleaning agents,

(3) gain a better understanding of the mechanisms occurring during membrane ageing,

(4) identify the most relevant membrane property capable to indicate these changes in membrane structural properties,

(5) assess the relative impact of fouling and cleaning on the filtration behaviour,

(6) assess current membrane cleaning protocols for polymeric microporous membranes during long-term progressive degradation of performance in parallel with membrane ageing and morphological changes and,

(7) characterise and possibly intervene the exact extent of the damage encountered by industrially-aged membranes during their lifetime.

2. MAIN OBJECTIVES AND OUTCOMES FROM INDIVIDUAL WORK PACKAGES

2.1 Critical assessment of the techniques characterising nascent changes in microporous membrane nature (work package 2 – see Appendix 1)

Various analytical tools are being used to observe and measure the mechanical, morphological and chemical effects of membrane ageing. The desirable characteristics of these methods generally include simplicity and lower cost. It is, however, important to use more sensitive and accurate methods jointly with novel and emerging technologies as it could allow the detection of nascent changes of the membrane. A thorough evaluation of the membrane characterisation techniques was executed by developing a ranking system to compare different criteria for the selection of the most suitable methods to assess membrane degradation or modification, as the result of chemical cleaning. The evaluation of the various analytical methods was reflected by the relative sensitivity of the tools in detecting the changes between new and aged flat sheet and hollow fibre polyvinylidene fluoride (PVDF) membranes. These membranes where statically aged in 2% NaOCl solution for the duration of one week. Clean water tests, contact angle, porometry, tensile test, Thermogravimetric analysis (TGA) and fourier transform infrared spectroscopy (FTIR) were completed on both flat sheet and hollow fibre membranes. Experimental reliability of the membrane characterisation techniques was also rigorously assessed for these membranes. The results from the sensitivity and reliably tests were used to rank all the tested characterisation techniques accordingly. Table 1 shows a summary of the recommended characterisation techniques for membrane ageing.

Table 1- Recommended characterisation techniques for membrane ageing analysis.

Techniques		Flat sheet	Hollow fibre
Clean water test		✓✓	✓✓
Porometry	Mean pore diameter	✗	✗
	Bubble pore diameter	✗	✓
Contact angle		✓✓	✗
Tensile test	Breaking force	✓	✗
	Breaking stress	✗	✗
	Ultimate elongation	✓	✓✓
	Young's Modulus (r2 = 0.90)	✗	✗
	Young's Modulus (r2 = 0.99)	✗	✓
TGA		✗	✓✓
FTIR		✓✓	✓✓

It was found that the various analytical techniques tested present different degrees of sensitivity and reliability towards the impact of the chemical agents. Therefore, the suitability of different membrane characterisation techniques as predictive tools for membrane ageing depends on membrane configuration (flat sheet/hollow fibre) and nature (material, average pore size). For example, clean water test was found to be more reliable for hollow fibre membranes, while contact angle was more suited for flat sheet membranes. Moreover, generally, different instruments operating range is limited, which makes them inappropriate for some samples. For instance, porometry analysis was only possible for membranes featuring pore size larger than 0.1 μm.

Finally, the relative contradiction observed between the testing methods indicated their limitations in measuring a single parameter. For example, many of the membrane characteristics (such as pore size, charge and roughness) can affect the contact angle value obtained from the goniometer. On the other hand, trends obtained from contact angle, membrane resistance and bubble point tests were successfully used to complement some proposed hypothesis regarding degradation mechanisms.

Findings in this section implied that in order to have a comprehensive record of the critical changes in membrane characteristics, more frequent sampling could be carried out in future work. Ageing tests at lower concentrations could also represent a better assessment of the nascent changes in the membrane properties in the early stages of degradation. Finally, characteristics such as: mean pore diameter, breaking force and ultimate elongation can also be suitable in complementing the results.

2.2 Static ageing for better understanding of degradation mechanisms (work package 3 – see Appendix 2)

From the knowledge gained during the literature review, it was observed that most research studies focused on effects of chemical agents on polystyrene/polyethersulfone (PS/PES), polypropylene (PP) and polyamide (PA) membranes. Very few studies have investigated the effects of cleaning chemicals on PVDF membranes, although this membrane material is becoming more widely used in the water industry. Additionally, limited information is available on the effects of all the main chemicals in the mixture of cleaning solutions in the commercial industry, namely caustics, citric acid and surfactants.

Four different cleaning agents (NaOCl, SDS, HCl and NaOH) were used to age five commercial porous membranes fabricated from various materials. The hollow fibre (HF) membranes used in this study included PVDF-HF, PP and PES. In addition, PVDF-hydrophilic (HPI) and –hydrophobic (HPO) flat sheet membranes were also used in the ageing tests. The maximum duration of the ageing experiment was 30 days, and membrane samples were analysed for a wide range of cleaning intensities. The results obtained from the membrane characterisation allowed (1) an arbitrary ranking of membrane stability against different chemical cleaning agents (2) a better understanding of the mechanisms occurring during membrane ageing. A summary of the assessment of the various membrane samples modification/degradation mechanisms by different cleaning agents is reported in Table 2.

Table 2- Summary of proposed degradation mechanisms by various cleaning agents for different membranes.

	NaOCl	SDS	HCl	NaOH
PVDF-HF	• Elimination of additives • Cross linking followed by chain scission	• Hydrophilisation	• Elimination of additives • Hydrophilisation	• Elimination of additives • Dehydrofluorination
PVDF-HPI	• Elimination of additives • Cross linking followed by chain scission	• Elimination of additives	• Elimination of additives • Hydrophilisation	• Elimination of additives • Dehydrofluorination
PVDF-HPO	• Cross linking	• More hydrophobic	-	• Cross linking
PP	• Elimination of additives • Chain scission	• Hydrophilisation • Swelling	• Hydrophilisation • Pore enhancement	• Hydrolysis • Chain scission
PES	• Chlorination of C-S bonds	• Hydrophilisation • Swelling	• Hydrophilisation	• Hydrolysis • Chain scission

"-": Not available.

The membrane ranking process indicated that PVDF-HF membrane properties were significantly modified after NaOCl and NaOH; whereas, SDS-aged membranes were not significantly affected. Both PVDF flat sheet membranes (PVDF-HPI and PVDF-HPO) were more sensitive to HCl exposure.

All oxidising agents (NaOCl, HCl and NaOH) generally resulted in the removal of membrane additives. For example, PVDF-HF membrane additives were removed by all three oxidising agents, although the extent of degradation of additives varied for different cleaning chemicals. NaOCl attack generally resulted in loss of membrane integrity through cross linking in PVDF, chain scission in PP, and chlorination of PES.

Cleaning with surfactants (SDS) led to subtle changes in membrane surface characteristics due to attachment of its hydrophobic tail on the membrane surface, which resulted in more hydrophilic and negatively-charged membrane. The membranes exposure to the inorganic acid seemed to only result in modification of membrane surface properties (hydrophilisation) without having a significant impact on membrane additives. Both caustic solutions resulted in degradation of all membranes core material.

2.3 Validity of accelerated ageing approach in terms of predicting membrane degradation in large-scale applications (work packages 4 & 5 – see appendices 3 and 4)

From earlier observations, it was proposed that ageing of the membrane could be partially responsible for the loss of hydraulic performances (i.e. increase in transmembrane pressure (TMP) during constant flux operation) observed during long-term operation [6]. A series of experiments was therefore especially designed to assess for the first time the relative impact of fouling and ageing on the changes of membrane performances. The membrane fouling/cleaning operation was conducted using a cross-flow tubular membrane module and a filtration set up, where all the equipment was pc-controlled. Due to time restrain, the chemical cleaning and rinsing time was artificially long and extensive, compared to the duration of the filtration. Feed solution was used to rinse the cleaning solution from the module after cleaning phase.

Sodium hypochlorite (NaOCl) was used during consecutive fouling and cleaning cyclical experiments. Following these experiments, a range of selected analytical techniques were carried out, allowing a complete assessment of the membrane state after ageing. Benchmark testings were also carried out in order to (1) identify the most relevant membrane property capable to indicate nascent changes in membrane chemical structure and (2) assess the relative impact of fouling and cleaning on the filtration behaviour.

Although the intensity of the chemical attack (up to 510,000 ppm.h) was generally lower than in the previous experiments (see work package 3 and [7]), degradation mechanisms of early stage of ageing was assessed from the current results. The apparent pore size enhancement of the PVDF membrane, obtained from resistance and total organic carbon (TOC) rejection measurements, was due to the combined effect of increase wetting and hydrophilicity of the membrane. Therefore, additives were expected to play a major role in the initial wetting of the membrane, while their presence during consecutive filtration is less crucial to suitable operation. Further characterisation of the various membrane samples indicated that NaOCl exposure lead to the potential damage of the PVDF membrane through cross linking. The fouling material used in these experiments potentially played a protective role for the membrane, as NaOCl was mainly used to oxidise the fouling rather than the membrane material.

Decline in PP membrane performances was significantly more pronounced, when compared to those obtained with the PVDF material. After filtration cleaning cycles, it was observed that the membrane resistance and bubble point pressure decreased significantly, indicating occurrence of pore enhancement. For PP membrane, the ageing mechanism potentially includes chain scission as a result of oxidisation of the polymer due to NaOCl attack. Moreover, the tensile properties of the PP membrane were also severely reduced after ageing.

In addition, the consecutive fouling/cleaning cyclical tests using NaOCl was conducted on two different industrially-aged PP membranes to assess their potential to sustain further ageing in a controlled (but accelerated) environment. By analysing the detailed characteristics of various membranes during accelerated ageing, it was expected that enough relevant information could be available to estimate the exact advancement of membrane ageing of an old membrane and to use those tests as benchmark for ageing assessment.

TGA analysis indicated that the industrially-aged membranes were relatively degraded, possibly through chain session during their long-term industrial use. The extent of the apparent degradation varied between the two samples. The initial assessment of the industrially-aged membranes via FTIR indicated that, even though the membrane additives still remained after the long-term industrial use of in the membrane material, different levels of membranes degradation (chain session) was evident considering the emerging bonds representing membrane degradation products.

After consecutive cyclical ageing, it was observed that both industrially-aged membranes featured similar hydraulic performances and presented a lower operating TMP compared to the new PP membranes, which can be due to the potential modification of these membranes properties during their lifetime. Although final TOC rejection of all aged membranes improved, formation of different types of fouling on the tested membranes was expected to be responsible for the observed variations in the TOC rejection performances obtained between the membranes.

Additionally, the consecutive cyclical ageing resulted in further degradation with observed decreased molecular weight of membrane material through oxidative degradation due to NaOCl attack. The observed embrittlement of these membranes further confirmed the occurrence of the oxidative degradation and chain scission. Indications of pore enhancement due to NaOCl exposure were also clearly observed.

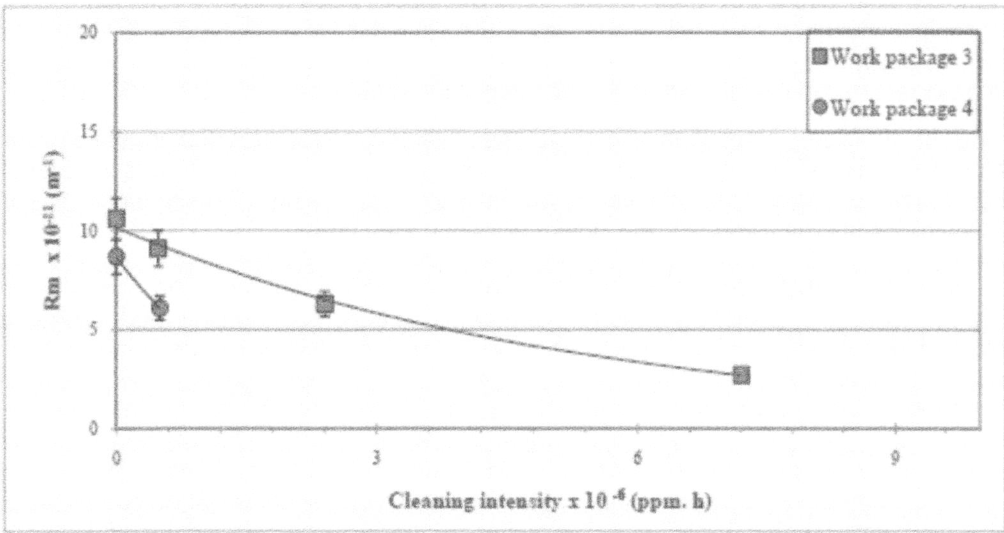

Figure 1- PVDF membrane resistance during static and consecutive ageing with 2% NaOCl.

The extent of membrane degradation during consecutive ageing was more advanced than the ones from static ageing tests. For instance, Figure 1 shows that the decreasing trend of membrane resistance (Rm) was more pronounced during the consecutive ageing study (MQ/NaOCl) compared to static ageing. This was probably due to the greater contact area between NaOCl and the membrane since the cleaning solution was briefly filtered through the membrane during the cleaning time. These observations were in agreement with previous study, in which both static and consecutive ageing tests were carried out using PVDF membrane and NaOCl [7]. In the later study, chemical modification of the membrane was observed to occur during the early stages of ageing, and extended contact time did not feature a significant impact on the membrane core material. Finally, the comparison of the statically-, consecutively- and industrially-aged membranes revealed that the effects of cleaning agent nature and concentration were significantly greater than the influence of ageing duration on membrane degradation.

3. CONCLUSIONS AND RECOMMENDATIONS

In this study, various aspects of membrane degradation were investigated through static and consecutive ageing tests conducted with different types of membrane and cleaning solutions. The results were also compared with industrially-aged membranes. It was postulated that the outcomes of accelerated ageing studies can be used as a tool to predict the loss of integrity of membranes in large-scale plants. However, it was observed that the extent of the apparent degradation varied due to samples history. This revealed a number of limitations to be addressed in terms of prediction of the large-scale membrane ageing using accelerated ageing trends; including:

1. The nature of the cleaning agent was observed to influence not only the mechanism of the membrane degradation, but also the severity of membrane integrity loss (work package #3)., since the cleaning chemicals applied in the industry were different to those used in this study, results can, therefore, be only compared to a limited extent.
2. The selection of the maximum concentration and duration that can be applied during accelerated ageing studies can create obstacles to match the cleaning intensities that are generally used in large-scale plants. Most of the previous accelerated ageing studies were based on high concentrations of cleaning agents in relatively short-term lab-scale experiments, with the aim to mimic long-term effect of membrane cleaning on membranes. On the other hand, the use of high concentrations of these chemicals has an instantaneous (sometimes annihilating) effect on some membranes or experimental apparatus. For example, relatively low concentration of NaOH in this study was applied due to instant reaction of PVDF when higher concentrations were used. Given the limited time available to conduct those accelerated ageing experiments, the overall cleaning intensity experienced in the industry can hardly been reached in the labs within a reasonable timeframe.
3. The lack of sufficient information regarding the cleaning protocols used in the industry, including the exact nature of the chemicals (mostly proprietary) and the frequency and the duration of chemical cleaning remains major limitation in this study. This indeed limits the evaluation of membrane ageing for the industrially-aged membranes.

Future studies focussing in more details on the effect of membrane degradation on product water quality is recommended. By assessing precisely the relative impact of fouling and ageing on membrane rejection performances, it is expected that the resulting improved understanding of the complex mechanisms occurring on the membrane surface could facilitate further validation of the technology for application in water and wastewater treatment and reuse. Additionally, frequent sampling from feed and permeate streams and further characterisation of the permeate samples through state-of-the-art analytical techniques like liquid chromatography-organic carbon detection (LC-OCD) and fluorescence emission-excitation matrix (FEEM), could also benefit result outcomes. In LC-OCD plots, the dissolved organic fractions in the water phase are characterised based on their molecular size, as well as charge and hydrophobicity, and the organic carbon in the sample is fractionised into five major fractions. The characterisation of the organic carbon in the aged membrane permeate could provide additional information on the non-retained species (hydrophilic/hydrophobic) and their relative organic fraction. Finally, the relative impact of physical stress on membrane ageing should also be assessed. Membranes are generally operated under relatively high turbulence conditions through aeration, liquid circulation, frequent drainage of tanks, backwashing operation. Notwithstanding the difficulties related to accelerate the impact of these operating conditions on lab scale, improvement in the design of the experimental plans could still be obtained for better assessment of the drop in mechanical strength of the membrane element.

4. REFERENCES

[1]. Kubota, N., T. Hasimoto, and Y. Mori, Microfiltration and Ultrafiltration, N.N. Li, et al., Editors.in Advanced Membrane Technology and Applications, John Wiley & Sons, New Jersey, 2008.

[2]. Arkhangelsky, E., D. Kuzmenko, N.V. Gitis, M. Vinogradov, S. Kuiry, and V. Gitis, Hypochlorite Cleaning Causes Degradation of Polymer Membranes Tribology Letters. 28 (2007) 109-116.

[3]. *BCC Research, Report Code: MST047B. Membrane and Separation Technology, Membrane Bioreactors: Global Markets* 2008.

[4]. Amjad, Z., Reverse osmosis: membrane technology, water chemistry & industrial applications, 1993.

[5]. Anil K. Pabby, S.S.H.R., Ana Maria Sastre, Handbook of membrane separations: chemical, pharmaceutical, food, and biotechnological applications, (2009).

[6]. Gaudichet-Maurin, E. and F. Thominette, Ageing of polysulfone ultrafiltration membranes in contact with bleach solutions, Journal of Membrane Science. 282 (2006) 198-204.

[7]. Puspitasari, V., A. Granville, P. Le-Clech, and V. Chen, Optimum Cleaning and Ageing Effect of Sodium Hypochlorite on Polyvinylidene Fluoride (PVDF) Membrane, (2010).

Appendix 1

Critical Assessment of the Techniques Characterising Nascent Changes in Microporous Membrane Nature Report No. 2

From the literature to the assessment of techniques reliability and sensitivity

Pierre LeClech

UNESCO Centre for Membrane Science and Technology, University of New South Wales

EXECUTIVE SUMMARY

Many microfiltration (MF) and ultrafiltration (UF) membranes are subjected to repetitive cycles of chemical cleaning as a fouling control strategy. The exposure to aggressive chemical cleaning agents could result in membrane degradation, leading to compromised process performances. Little is known on the nascent changes occurring at the molecular and structural level of a membrane in progressive ageing over prolonged periods. Adapting existing characterisation techniques to assess those changes is still required to better understand membrane integrity.

Accelerated ageing studies were conducted on polyvinylidene fluoride (PVDF) and polypropylene (PP) hollow fibre and PVDF flat sheet membranes with sodium hypochlorite (NaOCl). The concentration of NaOCl ranged between 4800 and 117600 ppmh. The ageing effects on both membrane types were observed by developing and evaluating the potential of a wide variety of membrane characterisation techniques including clean water and tensile tests, contact angle and porometry thermogravimetric (TGA) analysis. The reliability and sensitivity of the analytical techniques in assessing progressive membrane degradation were then assessed. The suitability of different membrane characterisation techniques as predictive tools for membrane ageing depends on membrane configuration and materials. The recommended techniques for flat sheet membranes are clean water test, contact angle analysis and tensile test. For hollow fibre membranes, clean water test, TGA and tensile test are proposed.

CONTENTS

FIGURES

TABLES

ABBREVIATIONS

AFM	- atomic force microscopy
ATR-FTIR	- attenuated total reflection - Fourier Transform infrared spectroscopy
CA	- cellulose acetate
CIL	- cleaning in line
CIP	- cleaning in place
CLSM	- confocal laser scanning microscopy
EDTA	- ethylene diamine tetra acetic acid
FS	- flat sheet
GPC	- gel permeation chromatography
HF	- hollow fiber
IC	- inorganic carbon
LMH	- Lm-2h-1
MBR	- membrane bioreactors
MF	- microfiltration
NaOCl	- sodium hypochlorite
PAN	- polyacrylontrile
PE	- polyethylene
PEG	- poly ethylene glycol
PES	- polyethersulfone
PP	- polypropylene
PS	- polysulfone
PVDF	- polyvinylidene fluoride
PVP	- polyvinylpyrrolidone
Rm	- membrane hydraulic resistance
SDS	- sodium dodecyl sulphate
SEM	- scanning electron microscopy
TC	- total carbon
TGA	- thermogravimetric analysis
TMP	- trans membrane pressure
TOC	- total organic carbon
UF	- ultrafiltration
WT	- water treatment
WWT	- wastewater treatment
XPS	- X-Ray photoelectron spectroscopy

1. INTRODUCTION

Microfiltration (MF) and ultrafiltration (UF) are microporous, low-pressure membrane processes that can be applied to remove microorganisms, suspended and colloidal particles. Separation is by sieving mechanism for microporous membranes, majorly governed by the pore diameter of the membrane. The pore size of these membranes typically varies between 1 nm-20 microns. Different types of membrane materials are in use for the separation applications, however polymer based membranes are mostly used in water treatment applications. A wide range of polymeric materials is available for membrane manufacture, varying in their physical, chemical and mechanical stability.

The demand for MF and UF in the water and wastewater treatment industry has increased greatly since the mid 1990s due to the requirement for technology to comply with new disinfection standards for drinking water and more stringent regulations for wastewater discharge respectively. This owes to the technology being used to replace conventional sand filtration in the clarification procedure (Kubota, Hasimoto et al. 2008), and requirements for advanced technology to comply with new disinfection standards for drinking water (Kennedy, Kamanyi et al. 2008). UF membranes have been reported to be able to combine high pathogen removal ability with minimal energy consumption (Arkhangelsky, Kuzmenko et al. 2007).

Application of MF/UF membranes in membrane bioreactors (MBR) is sharply increasing in the recent past years. A survey on MBR global market reports that this technology is expected to increase from $216.6 million in 2005 to $488 million in 2013 (2008). MBR have received most attention in the recent years due to their ability to combine biological treatment and filtration in one resulting in a compact process.

Application of MF/UF has also been considered as a promising alternative for the pretreatment to seawater desalination in the last ten years. MF/UF pre-treatment has various advantages over the conventional pre-treatment techniques, environmentally friendly in terms of limiting the sludge amount and use of chemicals, less foot print and cost effective.

During membrane filtration processes, fouling is an inevitable phenomenon and the general operating principle focuses on better managing this issue. The type and their degree of fouling depend on the feed water quality, pre-treatment, membrane properties and process design. Membrane fouling is the term used to describe the undesirable formation of deposits on membrane surfaces. Various membrane fouling types can be broadly classified as organic fouling, inorganic fouling, particulate deposition (colloidal) and biofouling (Amjad 1993). Fouling occurs when rejected solids are not transported from the surface of the membrane back to the bulk stream. This results in dissolved salts, suspended solids and microorganisms accumulating at the membrane surface. Some of the important mechanisms of membrane fouling are (Anil K. Pabby 2009), adsorption of feed components, clogging of pores, chemical interaction between solutes and membrane material, gel formation, and bacterial growth. In general fouling can be controlled by suitable pretreatment of the feed and membrane cleaning strategies.

Over long-term operation, membrane fouling is not completely reversible by simple physical cleaning (i.e. backwash and relaxation). Chemical cleaning is therefore performed periodically to restore the initial flux, and selection of the appropriate cleaning protocol is imperative. However, the major problem associated with chemical cleaning is the membrane degradation (or ageing) upon repeated fouling-cleaning cycles.

Although previous reports have reported possible chemical reactions between chemical agents and membrane, little is known on the nascent changes occurring at the molecular and structural level of membrane in progressive membrane ageing over prolonged period. The assessment of long-term progressive degradation of performance due to accumulated residual deposition has not been studied in parallel with membrane ageing and morphological changes. Adapting existing characterisation techniques to assess those changes is still required to better understand membrane integrity.

The present report aims to review the current understanding of membrane cleaning protocols for polymeric microporous membranes, its degradation during membrane cleaning and the techniques capable of detecting nascent changes of membrane material.

2. MEMBRANE MATERIALS

Membrane materials employed in the WT & WWT industries are mainly polymeric membranes due to their versatility, cost effectiveness and reliability. Polymeric materials generally used for UF/MF membranes are listed Table 2-1, in with their repeating units.

Table 2-1 Common polymeric materials used for MF/UF membrane manufacture.

Polymer	Repeating unit
polypropylene (PP)	$-[CH_2-CH(CH_3)]_n-$
polyethylene (PE)	$-[CH_2-CH_2]_n-$
polysulfone (PS)	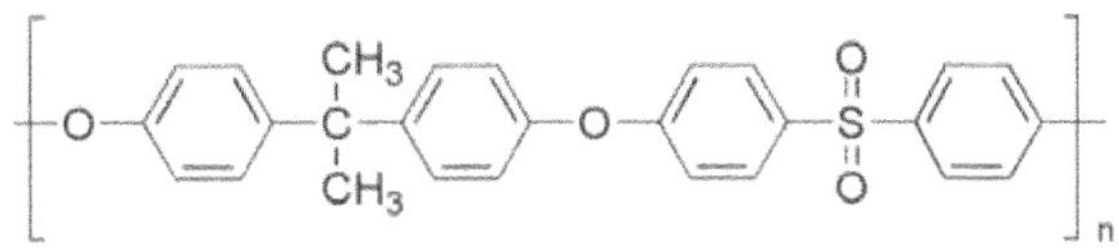
polyethersulfone (PES)	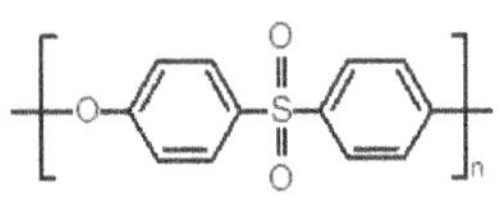
polyacrylontrile (PAN)	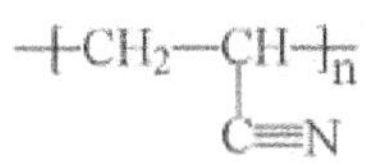
polyvinylidene fluoride (PVDF)	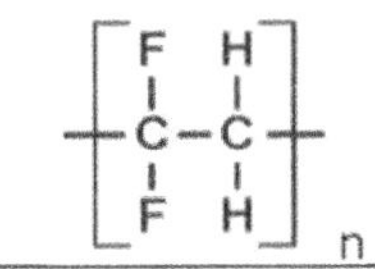

MBR takes a special position in the WT & WWT arena for its versatility. MBR does not need high pre-treatment steps, also capable producing good quality feed water for NF/RO in a single treatment step. Some of the commercially available MF/UF modules in MBRs are shown in Table 2-2.

Table 2-2 Commercialised MF/UF membranes for MBR.

Supplier	Material	Product name	Nominal pore size (μm)	Surface area (m^2)	Average flux (m/day)	References
Asahi Kasei	HF/PVDF	Microza®	0.1	25	0.8	(Hashimoto 2005; Asahi Kasei Chemicals 2006)
Zenon	HF/PVDF	ZW500C-D	0.04	31.6	0.9	(Judd and Judd 2006; Pearce 2008)
Mitsubishi-Rayon	HF/PE	SUR sterapore	0.4	15	0.4	(Euromba 2006)
Mitsubishi-Rayon	HF/PVDF	SADF	0.4	25	0.8	(Euromba 2006; Pearce 2008; Pearce 2008)
Memcor	HF/PVDF	Memjet B10R	0.04	10	0.6	(Euromba 2006; Pearce 2008; Pearce 2008)
Kubota	FS/PE	EK515	0.4	0.8	0.7	(Euromba 2006; Pearce 2008)

HF – hollow fiber, FS – flat sheet, PVDF – polyvinylidene fluoride, PE – polyethylene

From the membrane application viewpoint, characteristics of these membrane materials are compared based on three properties:

- Hydrophobic /hydrophilic interactions,
- Mechanical strength, and
- Chemical tolerance.

2.1 Hydrophobic/hydrophilic interactions

Membrane materials can be categorised based on their relative hydrophilicity. The degree of hydrophilicity or hydrophobicity impacts the wettability and fouling resistance of the membrane (Pearce 2007). Hydrophilic membranes are generally preferred for their higher fouling resistance and wettability, when compared to hydrophobic membranes. Hydrophilicity is also generally desired, as it allows stable and high filtration rate over long periods by inhibiting membrane fouling due to organic substances in raw water (Pearce 2007; Kubota, Hasimoto et al. 2008). In spite of the advantages, hydrophilic membranes feature comparatively poor robustness and durability. Most of the polymeric materials are naturally hydrophilic, whereas some other are made moderately hydrophilic through blending with additives and pore formers (Pearce 2007). Relative hydrophilicity of some of the common core polymeric membrane materials are shown in Figure 2-1.

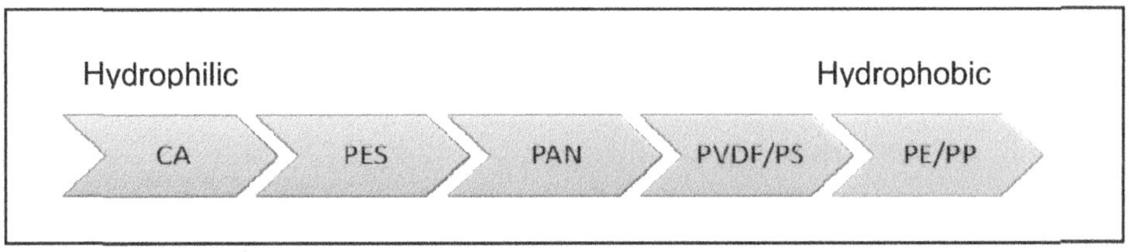

Figure 2-1 Relative hydrophilicity of polymeric membrane materials (Pearce 2007).

2.2 Mechanical stability

Membranes based on PS, CA and PP have been mainly used for water treatment industry in the '80s and '90s. Today, the most preferred membranes are made of PVDF, PS and PES; this is mainly attributed to the physical stability of the materials (Pearce 2007). Polymeric membranes with good mechanical properties are those which combine high burst and collapse strength with a reasonable degree of flexibility. Figure 2-2 shows how the commonly used polymers compare in these terms, with PS and PVDF being the strongest polymers. PVDF is significantly more flexible than all other materials, making it highly desirable if air scour is used for the cleaning process like in MBR systems.

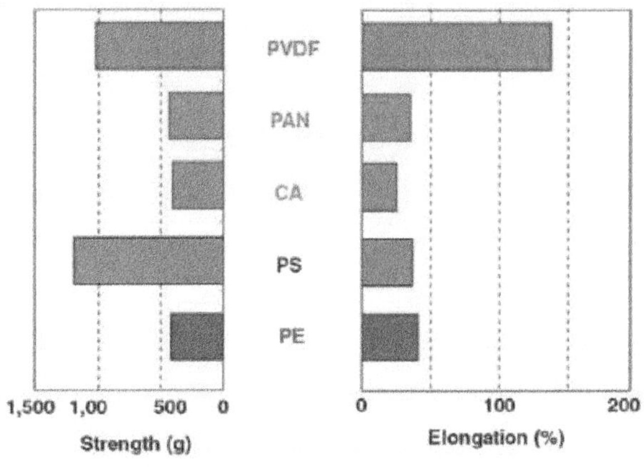

CA–cellulose acetate, PAN–polyacrylonitrile, PS–polysulfone, PVDF – polyvinylidene fluoride, PE – polyethylene

Figure 2-2 Comparison of physical stability of polymeric membrane materials

(Gijsbertsen-Abrahamse, Cornelissen et al. 2006).

2.3 Chemical tolerance

Desirable properties of membrane materials also include high degree tolerance to cleaning chemicals and withstand a wide range of operating temperatures and pH values. A good thermal resistance allow the use of elevated temperatures during cleaning, increasing its efficiencies. The chemical and thermal stability of polymers in general, and membranes in particular are determined by the structural factors of chain flexibility, chain interactions and crystallinity (Mulder 1996). A comparative account of the characteristics of common polymeric materials used in microporous membranes is presented in Table 2-3.

Table 2-3 Characteristics of microporous membrane materials

(adapted from Jacangelo and K.Noack 2005).

Polymeric material	Membrane type	Fouling resistance & cleanability	Mechanical strength	Operating pH range	Temperature limits	Oxidant resistance
Cellulosic	UF	Very good	-	4 - 8.5	Low	Moderate
PS	UF	Good	-	1 - 13	High	Good
PES	UF	Very good	-	2 - 12	High	Good
PAN	MF	Acceptable	Good	2 – 10.0	Moderate	Moderate
PP	MF	Acceptable	Good	2 - 14	Moderate	Poor
PVDF	MF/UF	Excellent	Good	2 – 10.5	Moderate	High

3. MEMBRANE FOULING AND CLEANING

Membrane fouling is an inevitable process occurring in membrane system resulting in loss of performances. Fouling is generally due to the deposition of materials on the membrane surface, or more importantly for MF and UF membranes, inside its porous structure (D'Souza and Mawson 2005). Some main categories of foulants are listed in Table 3-1.

Table 3-1 Types of foulants in water and wastewater treatments

(adapted from Mulder 1996; D'Souza and Mawson 2005; New Logic Research 2007).

Type	Characteristics	Examples	Common mode of fouling
Colloids	Particles 1 nm to 1 µm in size	Silica, clay and silt	Soiling
Organic	Carbon based, non living	Humic acids, oil, hydrocarbons, polymers	Soiling, absorption
Inorganic	Non-carbon based	Salts- $CaCO_3$, $CaSO_4$, $BaSO_4$. Metal hydroxides $Fe(OH)_x$, $Al(OH)_3$	Precipitation
Biological	Usually single celled organisms	Bacteria, algae, fungi	Biofilm formation and growth

In a full scale MBR operation, four different types of fouling rates has been explained by Kraume et al. (2009) as demonstrated in

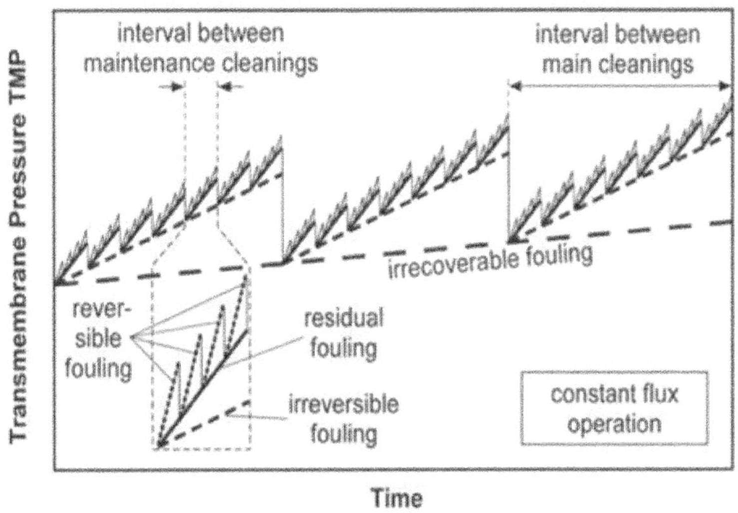

Figure 3-1 cake fouling, residual fouling, irreversible fouling and permanent/long-term fouling. Cake fouling and residual fouling is managed by regular maintenance cleaning also aiming at reducing the irreversible fouling. Irreversible fouling caused when the maintenance cleaning could not recover the flux and should be treated by chemical cleaning. Significant loss in permeability is observed after irreversible fouling and cleaning cycles, which is the irrecoverable fouling, developing over a long term operation.

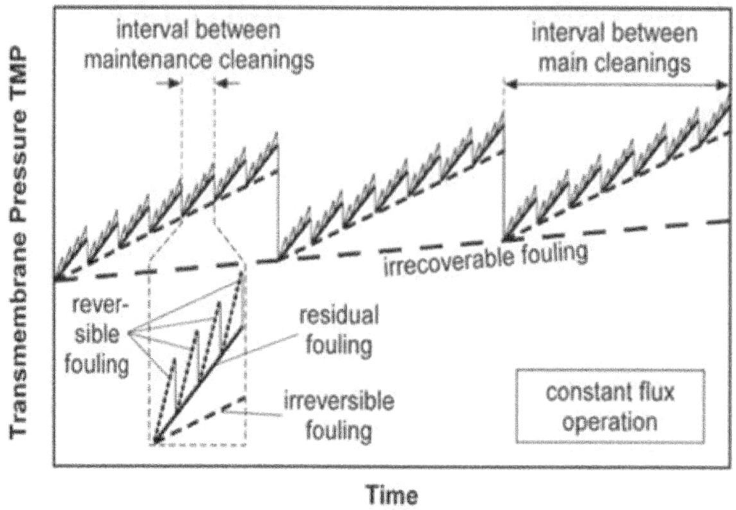

Figure 3-1 Different fouling rate schemes in full scale MBR operation (Kraume, Wedi et al. 2009)

Typical time intervals for different fouling rates are, Cake filtration - 10 min, Residual fouling - 1 to 2 weeks, Irreversible fouling - 6 to 12 months and Permanent/long-term irreversible fouling – several years.

3.1 Strategies for fouling control/removal

Fouling is affected by a wide variety of factors including membrane properties (such as pore size, pore size distribution, hydrophobicity and membrane material), process operating parameters (such as feed flow rate, temperature, pressure pH and concentration) and nature of components in the feed. As fouling occurs flux decline or TMP increase (for constant TMP or constant J operation respectively) generally results in negative impact on the economics of a given membrane operation (Mulder 1996).

Fouling control techniques include pre-treatment of the feed solution (prescreening chlorination or acidification), improvement in membrane properties (change in hydrophobicity and material), optimisation of operating conditions (increase fluid velocity, higher trans membrane pressure (TMP)) and cleaning.

3.1.1 Physical cleaning

Various methods to physically clean fouled membranes adopted for MF/UF during operation include relaxation, forward flush, backward flush and air flush (Lenntech 2009).

- Relaxation is the main physical cleaning technique in MBRs. During membrane relaxation, the filtration is paused resulting in back diffusion of the foulants.
- During forward flush cleaning, membranes are flushed with feed water or permeate in forward direction, flushing through the system more rapidly than during normal operation conditions. The rapid flow and turbulence loosens or detaches the particles absorbed or deposited to the membrane surface. However, the particles trapped inside the pores are not effectively removed.
- In the case of backward flushing, permeate is passed through the membrane (from permeate to feed side) at a high pressure, normally 2.5 times higher than the operating pressure. Pore-blocking particles are effectively removed by this process. In MBRs backwashing partially dislodge the loosely attached sludge cake from the membrane surface and remove most of the reversible fouling.

- Air flush works on the principle of forward flush, but air or air/water mixture is used for flushing the membrane. The air bubbles increase the turbulence and therefore efficient cleaning.

The above mentioned techniques generally help to regain the flux to an acceptable level, but when the initial flux is not restored; chemical cleaning needs to be performed. Chemical cleaning is often combined with flushing operation.

Membrane backwashing involves the pumping of permeate in a reverse direction through the membrane (similar to backwash flushing) allowing the removal of most of the reversible fouling due to pore blocking and partially dislodge loosely attached sludge cake from the membrane surface (Le-Clech, Chen et al. 2006).

Table 3-2 presents a summary of the effects of several fouling control measures in removing different types of foulants. Chemical cleaning are the highest positive effects in fouling control as the wide range of cleaning agents as they available would allow the removal of most types of foulants. Physical cleaning like backwashing would not be able to remove organic and inorganic foulants, while chlorination and acidification of the feed are ineffective in removing particulates.

Table 3-2 Effects of operating strategies on membrane fouling (Liu, Caothien et al. 2006).

Type of Fouling	**Effects of operating strategies**			
	Feed pre-treatment		**Cleaning**	
	Chlorination	Acidification	Physical	Chemical
Inorganic	-	+ +	-	+ +
Particulate	-	-	+ +	+ +
Microbial	+ +	+*	+	+ +
Organic	+	-	-	+ +

- no/negative effects, + some positive effects, ++ positive effects, * with feed chlorination

3.1.2 Chemical cleaning

Although the previously mentioned methods are capable to remove fouling to some extent, chemical cleaning is by far the most efficient strategy in the industry (Mulder 1996). Generally, membrane system operators follow a "10 percent rule"; i.e. the system is cleaned when the permeate rate drops or the differential pressure increases by 10% (New Logic Research 2007). Selection of appropriate cleaning protocol should be based on the membrane material, process configuration, operating conditions and nature of foulants.

Chemical cleaning is utilised when physical cleaning is not effective enough to restore the flux to an acceptable level (Kuzmenko, Arkhangelsky et al. 2005). Chemical reactions remove foulants by loosening, dissolving and keeping the foulants in dispersion and solution to prevent the re-fouling of already cleaned surface (Madaeni and Mansourpanah 2004). More details on the cleaning mechanisms are presented in Section 0. Despite the superiority in performance of chemical cleaning over other methods, the limitations include high cost of purchasing and disposing the chemical agents, waste issues and chemical degradation of the membrane.

3.2 Chemical cleaning in the WT & WWT industry

Currently, chemical cleaning practices are mostly based on recommendations from membrane manufacturers. Some of the manufacturers supply proprietary cleaners while others use commercial chemicals (Kennedy, Kamanyi et al. 2008). Although each membrane process requires unique cleaning procedure, some general formula/protocols and regimes of chemical cleaning do apply (D'Souza and Mawson 2005).

Different MF/UF chemical cleaning procedures are conducted at various frequencies; either daily, weekly or yearly. The cleaning types include chemically enhanced backwash (CEB), maintenance cleaning and intensive (recovery) cleaning. The objective of maintenance cleaning is to maintain design permeability and reduce the frequency of intensive cleaning. When further filtration is no longer sustainable due to elevated TMP or low flux, intensive cleaning is used.

Table 3-3 Examples of chemical cleaning protocol for MBR applications in water treatment.

Cleaning type	Frequency	Duration	Typical reagent	Concentra-tions (%)	References
Chemically enhanced backwash	Daily	10-15 min/ 4-6 h	NaOCl Citric acid/ NaOH	0.002 0.06	(Le-Clech, Chen et al. 2006; Alspach, Adham et al. 2008; Kennedy, Kamanyi et al. 2008)
Maintenance cleaning	Weekly	30 min/ cycle	NaOCl	0.01	(Le-Clech, Chen et al. 2006)
Intensive (recovery) cleaning	Once/twice per year		NaOCl	0.2-0.5	(Le-Clech, Chen et al. 2006)

For intensive cleaning, cleaning in line (CIL) and cleaning in place (CIP) methods are used. In CIL, chemical solutions are mainly back-flown under gravity inside the membrane. For CIP, the membrane tank is isolated and drained, the module is rinsed, soaked in cleaning solution and then the excess cleaning solution is washed. Some of the protocols used by MBR suppliers are included in Table 3-4.

Table 3-4 Intensive chemical cleaning protocols from MBR suppliers (Le-Clech, Fane et al. 2005).

	Chemicals	Concentra-tion (%)	Protocols	Cleaning duration (hours)
Mitsubishi	NaOCl Citric Acid	0.3 0.2	Backflush + soaking	2 + 2
Zenon	NaOCl Citric Acid	0.2 0.2	Backflush + recirculation	-
Memcor	NaOCl	0.01	Recirculation	-
Kubota	NaOCl Oxalic acid	0.5 1	Backflush + soaking	2

3.3 Cleaning agents and mechanisms

In practice, chemical cleaning is achieved through a variety of chemical and physical interactions between the cleaning solution and foulants (D'Souza and Mawson 2005). The main chemical reaction mechanisms between cleaning agents and foulants are hydrolysis, dissolution, dispersion, solubilisation, saponification, oxidation and chelation. The description of the mechanisms are summarised in Table 3-5.

Table 3-5 Chemical cleaning mechanisms.

Mechanism	Description	Reference
Hydrolysis	Decomposition of organic compounds by interaction with water	(Madaeni, Mohamamdi et al. 2001)
Dissolution	Formation of homogeneous mixture of solid/liquid with a solvent	(Madaeni, Mohamamdi et al. 2001)
Dispersion	Dispersion of particles in a continuous phase of a different composition	(Madaeni, Mohamamdi et al. 2001)
Solubilisation	To increase in the solubility of compounds (e.g. fat) in another liquid (water) using an agent (e.g. detergents)	(Liu, Caothien et al. 2006)
Saponification	Hydrolysis reaction of ester (e.g. fats) with alkali (e.g. NaOH) producing free alcohol and an acid salt	(D'Souza and Mawson 2005)
Oxidation	Gain of oxygen and/or loss of hydrogen of an organic substrate	(IUPAC 2009)
Chelation	Formation or presence of two or more separate bindings between a polydentate ligand and a single central atom	(IUPAC 2009)

Chemical cleaning agents are usually categorised according to their functional groups or the types of foulant they can typically remove. Some of the main chemicals used in the WT & WWT industries include alkali (caustics), oxidants and acids. Generally, oxidants are used first to remove organic matter, followed by organic acid coupled with mineral acid to remove metal hydroxides (Judd 2008) . The WT &WWT industries generally use hypochlorite as their preferred chemical cleaning agent due to its ready availability, reasonable price and high efficiency for both cleaning and bio-sanitation actions (Liu, Caothien et al. 2006; Pilutti and Nemeth 2006; Arkhangelsky, Kuzmenko et al. 2007). Some of the main categories are listed in Table 3-6 together with typical foulants removed, their major cleaning mechanisms and typical chemicals used.

Table 3-6 Main chemical cleaning agents used in WT & WWT and their properties

(adapted from (Trägårdh 1989; Shorrock and Bird 1998; Liu, Caothien et al. 2006).

Category	Typical Foulants	Mechanisms	Typical Chemicals
Caustics	Proteins, carbohydrates, fats	Hydrolysis, solubilisation, saponification	NaOH
Oxidants/ Disinfectants	Pathogenic microorganisms	Oxidation	NaOCl, H_2O_2, Peroxyacetic acid
Acids	Inorganic salts, oxide films	Solubilisation, hydrolysis	Citric, nitric, hydrochloric, phosphoric acid
Chelating Agents	Mineral deposits	Chelation	Citric acid, ethylenediaminetetraacetic acid (EDTA)
Surfactants	Fats, proteins, oil	Dispersion	detergents (anionics, non-ionics)
Enzymes	Proteins, lipids	Hydrolysis	Proteases, lipases

3.3.1 Caustics

Caustic compounds remove foulants mainly by solubilisation of fats and oils or by hydrolysis of organic materials such as polysaccharides and proteins. Water-soluble soap micelles react with fats and oils through the process of saponification (Liu, Caothien et al. 2006). Caustics cause the deprotonation of carboxylic and phenolic functional groups which increases the negative charge of humic substances resulting in an easier removal from the membrane surface (Kuzmenko, Arkhangelsky et al. 2005; Liu, Caothien et al. 2006). The repulsion between the negatively charged functional groups causes the organic substances to become stretched and linear, creating a loose fouling layer (Hong and Elimelech 1997). This layer allows better mass transfer of the cleaning agents to the inner portion of the fouling layer which increases the cleaning efficiency (Kennedy, Kamanyi et al. 2008). Hydroxide solutions are recommended to be used between pH 11 to 12 , as their high alkalinity also results in saponification, especially at warmer temperatures (>50°C) (Krack 1995; Cheryan 1998).

3.3.2 Oxidants

The introduction of strong chemical agents such as ClO^- or H_2O_2 result in the oxidation of organic compounds into more oxygen-containing functional groups such as ketones, aldehydes and carboxylic acids. These functional groups make the parent compounds more hydrophilic. At alkaline pH conditions, the increased amount of carboxylic groups on aromatic rings increases the negative charges of natural organic matter due to the dissociation of these acids, reducing adhesion of foulants to the membrane (Liu, Caothien et al. 2006).

3.3.3 Acids

Acids are used as chemical agents to remove scales, divalent cations and metal dioxides. The low pH also allows the hydrolysis of organic compounds such as polysaccharides and proteins. Citric acid especially is an effective agents as it dissolves iron oxide precipitates and other complexes (Liu, Caothien et al. 2006). It is also favoured over nitric acid due to its mildness, easy rinsing and its inability to corrode surfaces (D'Souza and Mawson 2005).

3.3.4 Chelating Agents

Chelating agent like ethylenediaminetetraacetic acid (EDTA) works by forming stable complex molecules with the foulant. After the complex molecules are formed, they detach from the membrane and could be easily removed. Chelating agents prevents the re-deposition and/or removal of mineral deposits. Chelating agents are used in soaps to remove metallic contaminant due to its good ability to combine with metals (Kazemimoghadam and Mohammadi 2007).

3.3.5 Surfactants

Surfactants play various roles in membrane cleaning. They help in bringing the detergent and soil together by surface wetting, facilitate foulant removal, assist in solubilisation and emulsification, stabilize foam, and adsorb on surfaces to amend properties of the surface (Cox 1994; D'Souza and Mawson 2005). By altering the interfacial tension of water, better separation of deposited materials from the membrane surface is observed (Kazemimoghadam and Mohammadi 2007). Having both hydrophobic and hydrophilic structures, surfactants form micelles with fats, proteins and oil in water. Surfactants also work as biocides by disrupting the functions of bacterial cell walls (Liu, Caothien et al. 2006).

3.3.6 Enzymes

Enzymatic detergents are usually used in the event of high soiling or in systems with pH limit of 10. Enzymes work by scissoring specific functional groups in the protein strands, with maximum efficiency at an optimum pH and enzyme concentration (D'Souza and Mawson 2005). The advantages of enzymes over other chemical agents include biodegradability, less destructive to the membranes, lower temperature operation (and hence lower operating cost) and being highly substrate and reaction specific. All of these could lengthen membrane lifespan (Argüello, Álvarez et al. 2003; D'Souza and

Mawson 2005). Potential drawbacks include narrow optimum pH operating range, relatively higher cost than conventional cleaning agents, and the relatively slower rate of enzyme activity (D'Souza and Mawson 2005).

3.4 Combined cleaning effects

Although single cleaning agent is still widely used in the industry, formulated detergents are more commonly proposed with mixtures, typically consisting of caustics, chelating agents and surfactant (D'Souza and Mawson 2005). Enhanced efficiencies of the cleaning process through combined cleaning agents are reported from previous works (Table 3-7).

Table 3-7 Combined effects of chemical cleaning agents.

Combined recipe	Details	Reference
NaOH + EDTA +SDS	NaOH changes pH – more efficient effects of EDTA + SDS	(Kazemimoghadam and Mohammadi 2007)
NaOH + detergents	Enhanced cleaning effects	(Maartens, Swart et al. 1998)
NaOH + oxidants	Enchanced cleaning efficiency- fouling layer have a more open structure in caustic conditions. This allows chlorine to reach the inner layer of foulant materials	(Kennedy, Kamanyi et al. 2008)
NaOH + NaOCl	Enchanced cleaning effects	(Krack 1995)
Enzymes + alkali	Effective in severe membrane fouling when protein foulants are concentrated to a very high solid content	(Liu, Caothien et al. 2006)

SDS – Sodium dodecyl sulfate, EDTA – ethylenediaminetetraacetic acid

3.5 Cleaning strategies

The main factors to consider for an effective membrane cleaning regime include the cleaning agents used, the feed characteristics and foulant types to be removed, operating conditions, cleaning frequency and duration, and membrane materials .

The important factors for selection of cleaning agent include its optimum concentration, its chemical stability during use, costs and safety. The chemical concentrations could affect the equilibrium and the rates of reaction in the cleaning process. Higher dosages of cleaning agents significantly increase diffusion of the chemicals to near-membrane surface layers for faster and more efficient CIP in a short period of time (Mohammadi, Madaeni et al. 2003; Kuzmenko, Arkhangelsky et al. 2005). Previous studies have shown that high concentrations of free chlorine and NaOH (0.1%) resulted in an almost complete restoration of the initial flux (Hong and Elimelech 1997; Kuzmenko, Arkhangelsky et al. 2005). However, higher concentration did not increase the performance (Madaeni and Mansourpanah 2004) and in some cases decrease the cleaning efficiency (Madaeni, Mohamamdi et al. 2001; Li, Li et al. 2005; Puspitasari, Granville et al. 2010).

Operating conditions: Chemical cleaning effectiveness is also affected by operating conditions such as cross flow velocity, turbulence in the vicinity of the membrane surface, temperature, pH and cleaning time (Madaeni, Mohamamdi et al. 2001). Temperature affects cleaning by changing the solubility of the foulants or cleaning agent and shifting the reaction equilibrium (Kennedy, Kamanyi et al. 2008). In general, increased temperature results in higher reaction kinetics which amplifies the cleaning rate. However, some membrane materials could undergo thermal degradation if the temperature is too high. Most of the polymeric membranes used in the industry could be operated at temperature higher than 40°C, except for CA which is limited to 35°C (Pearce 2007).

Cleaning sequence frequency and durations: The sequence of cleaning agents applied is also important. Most cleaning regimes follow alkali with acid steps with rinsing in between (D'Souza and Mawson 2005). This is due to the fact that acid cleaning is not capable to hydrolyse and solubilise the protective layer around microorganisms. Acids instead "tighten" the biofilm, which can cause further performance decline of the membrane (Johnson 2009). Generally longer duration would result in better cleaning, until an optimum period is achieved.

3.6 Cleaning efficiencies

A compilation of recent studies conducted on cleaning efficiencies of different chemical agents are summarised in Table 3-8.

Table 3-8 Cleaning efficiencies of different chemical agents.

Cleaning agent	Membrane system	Optimum operating conditions	Feed	Cleaning efficiencies (%)	Reference
NaOCl	UF-PES	0.01-0.5%	bovine serum albumin (BSA)	100	(Kuzmenko, Arkhangelsky et al. 2005)
NaOH	MF-PVDF	0.2%, 50°C	α-lactalbumin	90	(Bansal, Al-Ali et al. 2006)
SDS	UF-PS	0.3%, 42°C	fermentation broth of glutamic acid	94	(Li, Li et al. 2005)
NaOCl + protease detergent + NaOH	UF-PS	0.015%, 50°C	soy flour	90	(Sayed Razavi, Harris et al. 1996)
NaOCl + NaOH + HNO_3	MF-PVDF	0.05%	biofoulants	77	(Lim and Bai 2003; Puspitasari, Granville et al. 2010)

UF - Ultrafiltration, MF – Microfiltration, SDS – Sodium dodecyl sulfate, PES – Polyethersulfone, PVDF – polyvinylidene fluoride, PS – Polysulfone

A study by Kimura *et al.* (2004) compared the effects of individual agent and combinations. In this study, As seen in Figure 3-2, the best performing individual chemical agent is NaOCl (500 ppm) and the best combination is oxalic acid (pH 2) and NaOCl (500 ppm) followed by NaOH/NaOCl and HCl (pH 2)/NaOCl. As expected, these three combinations work better than any individual cleaner (Kimura, Hane et al.).

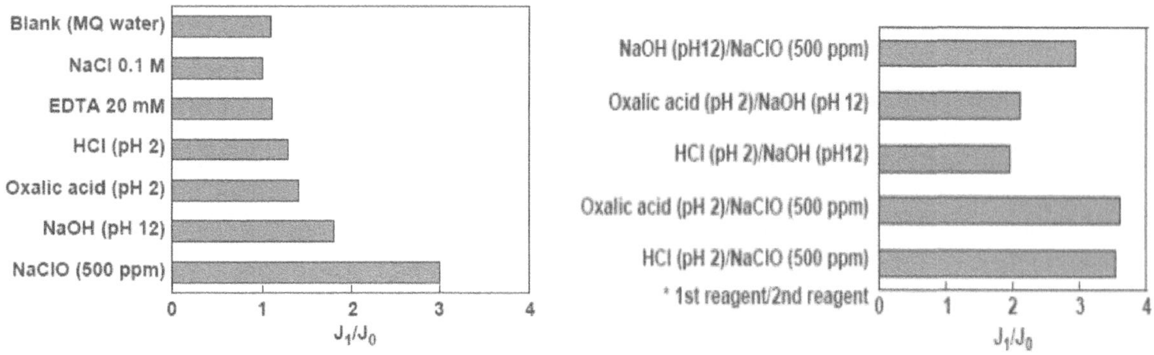

(a) (b)

J_1 – Pure water flux after chemical cleaning, J_0 – Pure water flux before chemical cleaning

Figure 3-2 Flux recovery for single (a) and combined chemical agents (b)

(Kimura, Hane et al. 2004).

4. POLYMERIC MEMBRANES AGEING MECHANISMS

4.1 Membrane degradation by cleaning agents

Chemical cleaning is routinely performed with corrosive and strong oxidizing agents. Many recent studies however clearly indicated that the permeability performances of MF and UF membranes deteriorate upon chemical cleaning, signalling potential failure of the membranes (Baah and Baah 2002; Childress, Le-Clech et al. 2005; Gaudichet-Maurin and Thominette 2006; Kwon and Leckie 2006; Arkhangelsky, Kuzmenko et al. 2007; Causserand, Rouaix et al. 2008).

Membrane integrity failures result in high economic and environmental costs from frequent membrane replacements. Health concerns are also a concern, where a single breach in the membrane fibre could compromise the removal of bacteria and viruses in the water treatment industry (Gitis, Haught et al. 2006). The loss of membrane integrity is a major problem in continuous flow submerged hollow fibre MBR system as solids and particles would be allowed to quickly clog the broken fibre and increase passage of high concentrations of biomass to the permeate (Le-Clech, Fane et al. 2005). Backwash efficiency can also be reduced, as the solids are be forced out of the fibre and the compromised membrane will need to be replaced (Le-Clech, Fane et al. 2005).

Information on membrane life time is limited, but various researches have reported significant ageing from the extended contact between membrane and cleaning solutions. Anecdotal evidences have suggested that membrane deterioration is accelerated by excessive cleaning with oxidative chemicals (Le-Clech, Fane et al. 2005). Hypochlorite solutions for example, react rapidly and unselectively with organic substances and would therefore attack the polymeric materials (Gaudichet-Maurin and Thominette 2006).

PES membranes are considered to be tolerant to oxidants, up to a chlorine concentration of 250 000 mg.L, and capable of operating between a wide range of pH, 2 and 12 (Arkhangelsky, Kuzmenko et al. 2007). Impact of hypochlorite cleaning on UF membrane has been compiled in Table 4-1.

Table 4-1 Impact of hypochlorite cleaning on UF membranes.

Membrane material	Impact reported	Reference
PES/PVP	Chain scission in the PVP occurs, with approximately 1% of pyrrolidine ring opening. The molecular weight of PVP decreases, especially at pH 11.5	(Wienk, Meuleman et al. 1995)
PS/PVP	High hypochlorite concentration make chain scission of PS molecule, consequently affects membrane texture, mechanical properties.	(Rouaix, Causserand et al. 2006)
PS	Decrease in elastic elongation with exposure. Generates an embrittlement of the fibre, occurs by chain scission with the hydroxyl radical OH^{-}.	(Gaudichet-Maurin and Thominette 2006; Thominette, Farnault et al. 2006)
PES	Effective restoration of the initial flux but severe fouling after every cleaning.	(Kuzmenko, Arkhangelsky et al. 2005)
PES	Effective removal of fouling, but resulted in increase in severe fouling and increased electronegativity of the cleaned membrane. The membrane integrity was lost at high hypochlorite concentrations.	(Arkhangelsky, Kuzmenko et al. 2007)
PS/PVP	Chain breakage of polysulfone, the degradation mechanism involves radical oxidation by hydroxyl radicals. Radical oxidation can be catalysed by the presence of metal ions like Fe^{3+} and Cu^{2+}; however can be inhibited by an antiradical (inhibitor) like tertiobutanol, added to hypochlorite solution during cleaning.	(Causserand, Rouaix et al. 2008)

PES – polyethersulfone, PVP – polyvinylpyrrolidone, PS – polysufone,

4.2 Polyvinylidene fluoride (PVDF)

PVDF polymer is recognised as a strong, robust material which allows extended membrane lifetime, many WT and WWT plants currently rely on the use of PVDF based membrane processes.

Limited literature dealing with the effect of cleaning agents on PVDF membranes is available. Under accelerated cleaning conditions (2% NaOCl), FTIR analysis indicated possible ageing effect from the chemical changes in the membrane functional groups. The same study also revealed a two-step degradation mechanism for the polymer. The surface modification substance was first removed, followed by an increase in the membrane's relative hydrophilicity (Puspitasari, Granville et al. 2010).

Studies of PVDF ageing from applications other than membrane could also help in identifying possible causes of polymer degradation. Ageing results from photovoltaic cells encapsulation shows that physical ageing occurs on the PVDF material used (Oreski and Wallner 2005). A summary of the various studies conducted on PVDF degradation is shown in Table 4-2.

Table 4-2 Previous studies on PVDF polymer ageing.

Application	Chemical Agent	Operating Conditions	Autopsy techniques	Results	Reference
MF-FS	NaOCl	2%,	FTIR, XPS, FESEM, Contact Angle	PVP removal, increased hydrophilicity	(Puspitasari, Granville et al. 2010)
Photovoltaic cells encapsulation	-	85°C, 1000h	ATR-FTIR, Tensile tests	Physical ageing- Increased yield stress	(Oreski and Wallner 2005)
MF-HF	NaOCl	0.5%, 25°C, 2 months	Tensile tests	No degradation	(Kubota, Hasimoto et al. 2008)
MF-HF	NaOH	4%, 25°C, 2 months	Tensile tests	No degradation	(Kubota, Hasimoto et al. 2008)

MF – microfiltration, FS – flat sheet, HF – hollow fiber, ATR-FTIR – Attenuated total reflection - fourier transform infrared spectroscopy, XPS – X-ray photoelectron spectroscopy, FESEM – Field electron scanning microscopy, PVP – polyvinylpyrrolidone

4.3 Ageing of Polysulfone (PS) / Polyethersulfone (PES) membrane

Following NaOCl treatment, PS and PES membranes have been reported to experience pore enlargement and lower elongation propensity for (Wolff and Zydney 2004; Gaudichet-Maurin and Thominette 2006; Rouaix, Causserand et al. 2006), changes in functional groups, chlorination and changes in hydrophilicity (Kwon and Leckie 2006; Arkhangelsky, Kuzmenko et al. 2007). XPS analysis on PES membrane cleaned with NaOCl showed an increase in carbon and sulphur content and decreased amount of oxygen. The elevated concentrations of sulphur and carbon in the upper part of the PES membrane was proposed to indicate partial scission of the sulfonyl Ph–S bond, as seen in Figure 4-1. The reaction occurs through an intermediate step in which phenyl sulfinate $PhSO_2$ dissociates into $PhSO_2^-$ followed by the formation of $PhSO_3^-$ (Arkhangelsky, Kuzmenko et al. 2007).

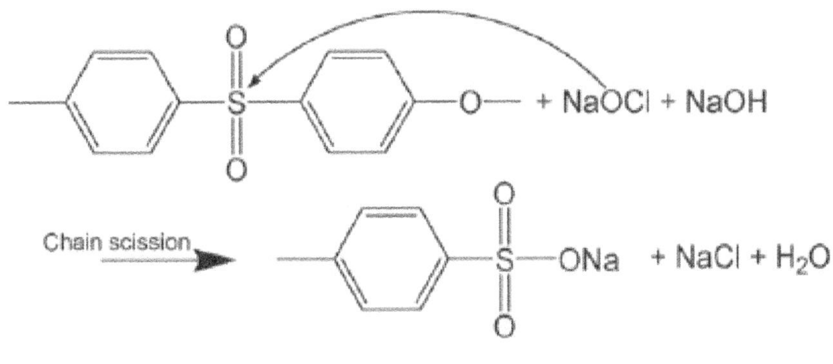

Figure 4-1 Proposed NaOCl attack mechanism on PES membrane

(Arkhangelsky, Kuzmenko et al. 2007).

Exposure to NaOCl at high concentration and extended time was found to change the mechanical properties of the membrane, such as the elongation and tensile strength at break point (Rouaix, Causserand et al. 2006). The changes in mechanical properties are supported by observed changes in the morphology of the PS membrane under study.

A study conducted by Mohammadi *et al.* (2003) on PS membranes showed that a combination of SDS + NaOH are the best agents in removing microorganisms, proteins, fats or minerals, followed by NaOCl + NaOH. The same study also observed that NaOCl membrane cleaning above 0.4% resulted in flux recovery of more than 100%, which clearly indicated significant changes in membrane characteristics and possibly its integrity. The results were attributed to membrane structure damage from contact with high concentration NaOCl or increased hydrophilicity of the membrane surface, which resulted in higher permeability (Mohammadi, Madaeni et al. 2003).

4.4 Attrition of surface modifiers

PS, PES, PAN and PVDF membranes are naturally hydrophobic and must be blended with additives and pore-formers to increase their hydrophilicity. The most popular additive in use is polyvinylpyrrolidone (PVP).

The ageing of membrane additives such as PVP in bleach solution has been extensively studied (Gaudichet-Maurin and Thominette 2006). In a study conducted on modified membrane, two distinct mechanisms of degradation of PVP from bleach solutions have been reported: (1) chain scission resulting from attack by hydroxyl radicals formed in acidic media (Figure 4-2) and (2) pyrrolidone ring opening by ionic process in alkaline media (Roesink 1989). XPS analysis of PS membrane modified with PVP showed the formation of COO^- functional group originating from the PVP after 4% NaOCl treatment. This was supported by the reduction of PVP concentration by 30%, indicated by the significantly lower nitrogen content originating from the PVP (Wolff and Zydney 2004).

Polyethylene glycol (PEG) is also often used as pore-former in polymeric membranes can undergo radical degradation induced by hydroxyl radicals, with formation of aldehydic chain ends as shown in Figure 4-3 (Fukatsu and Kokot 2001).

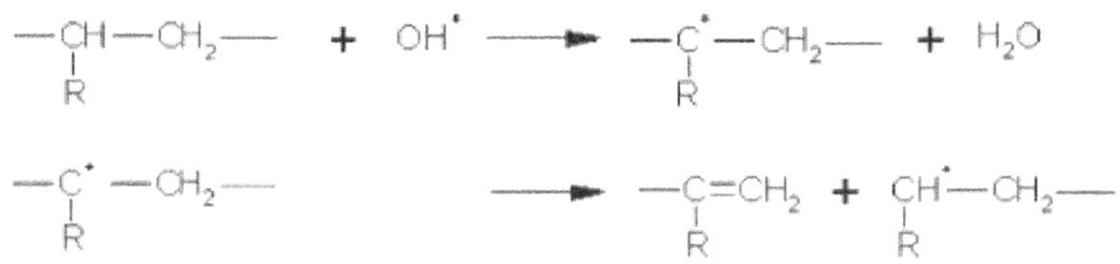

Figure 4-2 PVP chain scission mechanism (Roesink 1989).

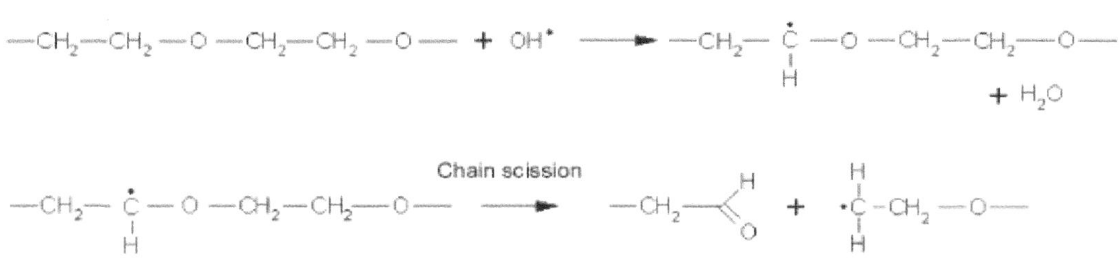

Figure 4-3 PEG chain scission mechanism (Fukatsu and Kokot 2001).

4.5 Other parameters affecting membrane ageing

Effect of temperature on the extent of damage was studied for PS membrane, with 100 ppm hypochlorite concentration, pH 7, contact time 16 days and temperature 6, 25 and 40°C. As expected, rise in temperature enhances the fibre weakening. At 6°C, there was no appreciable loss in tensile strength whereas 20% elongation loss was observed, after 8 days of immersion. At 40°C, elongation loss of 80% was seen after two days of immersion, and after two days tensile test and elongation loss could not be determined, since the membrane turned out brittle. At room temperature, 25°C, the same elongation loss was reached after four days. The surface morphology, filamentous was retained for 6°C even after 16 days of immersion. Also at higher temperatures, greatest decrease in molar masses was observed, indicating the maximum polymer chain scission (Rouaix, Causserand et al. 2006). Thus, at elevated temperatures the cleaning efficiency is high; consequently chance for membrane damage is studied to be high. A careful selection of temperature will help in attaining optimum cleaning efficiency and membrane damage.

The impact of hypochlorite was found to be catalysed due in the presence of metal ions like Fe^{3+} and Cu^{2+} in the interaction between hypochlorite and PS membranes, thus the membrane damage could be intense and fast. However, the simultaneous presence of these two ions seems to have no additive or synergistic effect. On the other hand, presence of tertiobutanol inhibits the oxidation reaction, when added with the hypochlorite solution during cleaning. The inhibitors act on the hydroxyl radical and arrest the membrane degradation mechanism involving radical oxidation (Causserand, Rouaix et al. 2008).

5. PREDICTIVE TOOLS ASSESSING MEMBRANE AGEING

Some of the testing protocols/analytical techniques employed for detecting membrane ageing are summarised in Figure 5-1. These techniques are complementary and could help to develop an understanding of the membrane ageing phenomenon.

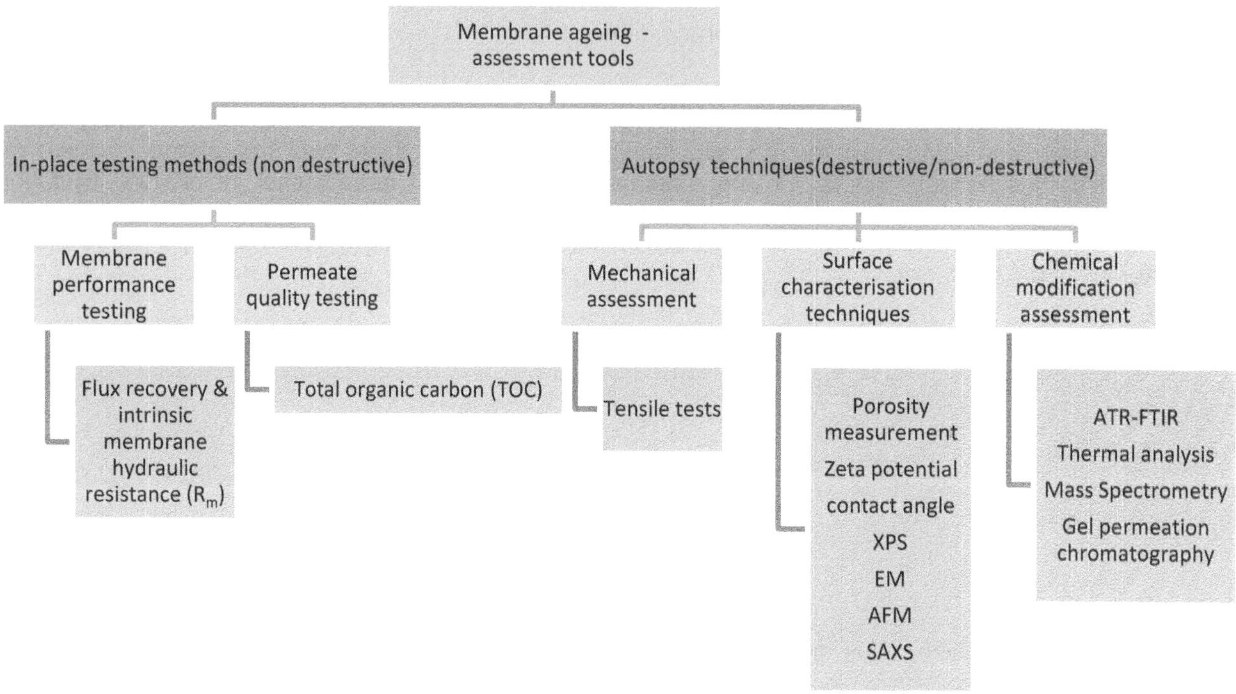

Figure 5-1 Assessment tools for membrane ageing.

5.1 In-place testing methods

In-place testing techniques are directly performed on the membrane modules. The testing may involve the assessment of the membrane hydraulic and rejection performances and could be calculated during the membrane filtration itself, without damaging the membrane surface, neither disturbing the separation process.

5.1.1 Flux recovery and intrinsic membrane hydraulic resistance (Rm)

The effect of membrane ageing due to chemical cleaning on membrane permeability/filterability could be assessed by measuring the intrinsic membrane hydraulic resistance (R_m) before and after ageing. In bench scale the membrane ageing could be assessed in two ways: (1) Studying the membrane resistance before fouling, after fouling and post-chemical cleaning. In this case, the flux recovery will not be 100% due to the irrecoverable fouling. (2) Studying the membrane ageing through accelerated chemical cleaning wherein no fouling would be performed; stability of membrane towards the cleaning agents will be directly assessed.

An ageing study assessing PVDF MF membranes had shown that after fouling and chemical cleaning, the initial permeability of the tested membrane was only partially recovered. The flux recovery seemed to be more efficient when applied to the PS MF membranes (Pontié, Thekkedath et al. 2007). The degree of cleaning efficiency of different chemicals was evaluated for BSA fouled PS membrane through permeability studies (Kuzmenko, Arkhangelsky et al. 2005). The percent flux recovery was calculated by 30 min filtration of deionised water after BSA membrane filtration and cleaning-in-place operation. The ratio of specific flux, in terms of $L/h.m^2.bar$ at 20°C of fouled or cleaned membrane (J) to the fresh membranes initial flux (J_0) was estimated as indicated in Equation 1. The flux recovery obtained with different cleaning agents is presented in Figure 5–2.

$$\text{Flux recovery (\%)} = \left(\frac{J}{J_0}\right)_{\text{DI water}} \times 100 \qquad \textit{Equation } 1$$

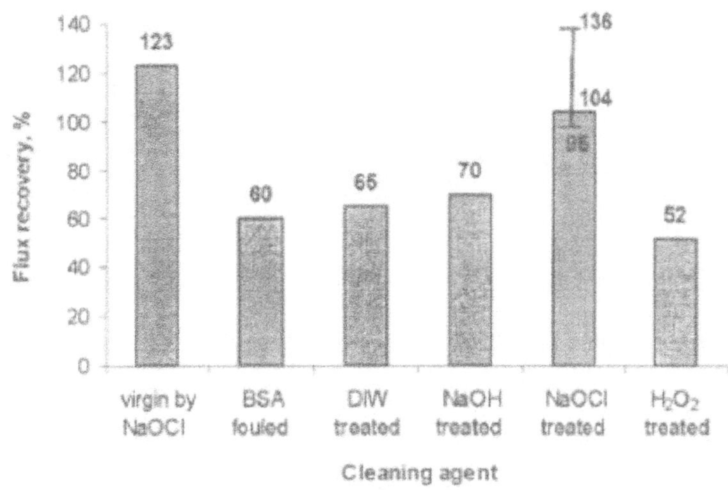

Figure 5-2 Percent flux recovery of membrane permeability with different cleaning agents, for BSA fouled PS membrane

(Kuzmenko, Arkhangelsky et al. 2005).

Cleaning with deionised water resulted in flux gain of 5%, followed by NaOH with 10% gain. The ultimate cleaning agent in this study is hypochlorite with maximum flux recovery. Cleaning with hydrogen peroxide actually minimised the flux, which could be due to the swelling of membrane or formation of some gel like products blocking the membrane pores (Kuzmenko, Arkhangelsky et al. 2005).

A similar study was performed recently by Arnal *et al.* (2009) on PS UF membranes used for treating surface water. The degree of flux recovery was compared to assess the effectiveness of the different cleaning agents in removing fouling and recovering membrane performance. As seen in Figure 5-3, the best performing chemical agent is the commercial solution Auxiclean B.13. Cleaning with NaOCl and H_2O_2 showed relatively acceptable flux recovery at 25^{0}C. However, at higher temperature NaOCl did not show any significant effect while H_2O_2 showed improvement in permeability recovery (Arnal, Garcia-Fayos et al. 2009).

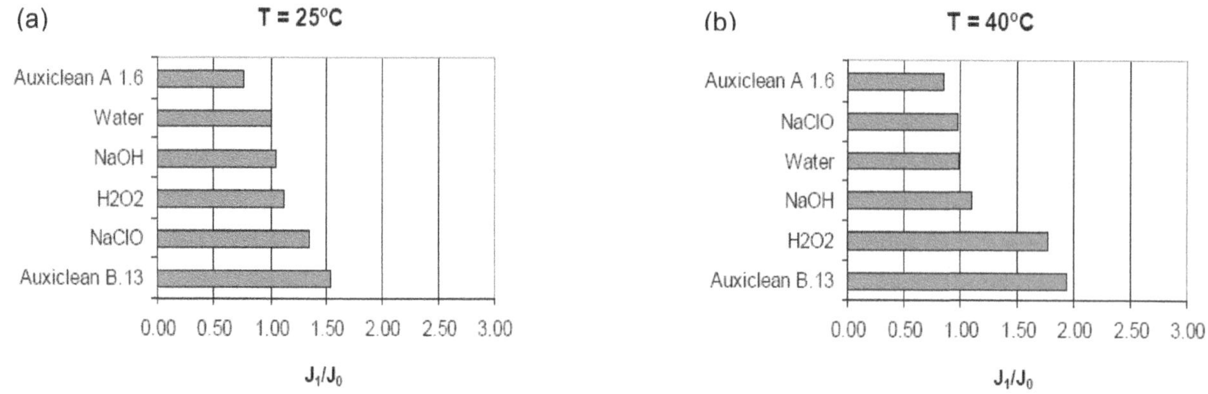

J_1 – Pure water flux after chemical cleaning, J_0 – Pure water flux before chemical cleaning

Figure 5-3 Flux recovery of PS membrane at (a) 25^{0}C and (b) 40^{0}C

(Arnal, Garcia-Fayos et al. 2009).

5.1.2 Permeate quality testing

To assess the membrane ageing from the permeate quality, TOC measurement could be used as outlined in the ASTM D6908 – 06 Standard Practice for Integrity Testing of Water Filtration Membrane Systems. Zhang and Liu (2003) investigated the effect of membrane cleaning on HF blend membranes of PS and dichlorodiphenyl sulfone–phenolphthalein (PDC) condensation polymer fouled by sodium hydroxide, calcium chloride and Turkey red oil, an anion surfactant contained in banknote printing wastewater. The cleaning procedure involved HCl cleaning prior to NaOH. As shown in Figure 5-4, the amount of TOC concentrations in the NaOH cleaning solutions was significantly higher compared with HCl cleaning solution suggesting that most organic foulants were removed during NaOH cleaning. During HCl solution cleaning step, a denser gel layer formed on top of the surface of the membrane. However, HCl solution helps transforming the foulants into forms that can be destroyed with NaOH and transformed into soluble compounds (Zhang and Liu 2003)

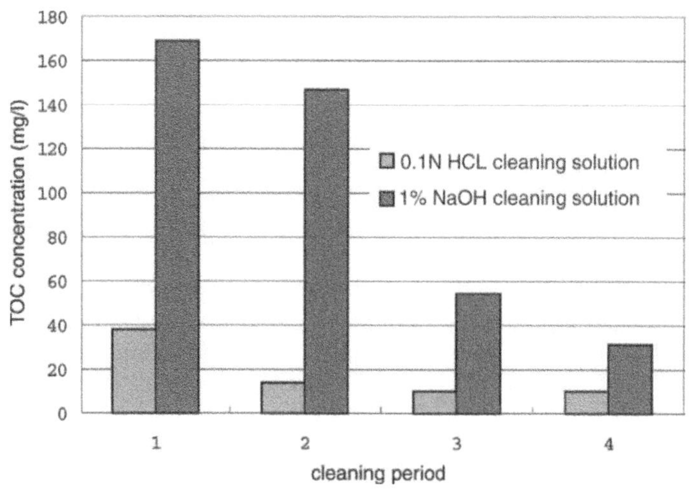

Figure 5-4 TOC of different cleaning solution on PS/PDC HF blend membrane

(Zhang and Liu 2003).

5.2 Mechanical assessment

The strength of membrane materials play a very important role in membrane breakage (Gijsbertsen-Abrahamse, Cornelissen et al. 2006). It has been reported that chain breaking from chemical cleaning induces a gradual deformation of the membrane skin layer and deterioration in mechanical strength of UF membranes (Arkhangelsky, Kuzmenko et al. 2007). Gaudichet-Maurin *et al.* (2006) reported a decrease in elastic elongation of PES membranes exposed to hypochlorite solution. There is, however, still limited information on the mechanical properties like polymer elongation and tensile strength in UF/MF membranes under constant operation and repeated cleaning (Rouaix, Causserand et al. 2006).

Tensile test

The main purpose of performing tensile test on polymeric membrane is to measure the material strength after being subjected to chemical cleaning. Previous reports have shown a correlation between structural changes of a membrane after cleaning and the mechanical strength after deterioration.(Gaudichet-Maurin and Thominette 2006).

In a tensile test, the membrane, with a known length, is clamped, and force is applied at a constant rate to draw each end apart. The membrane elongates and eventually breaks. The applied force over distance is recorded until the membrane breaks, resulting in force versus extension curve. This curve is converted into stress-strain curve (

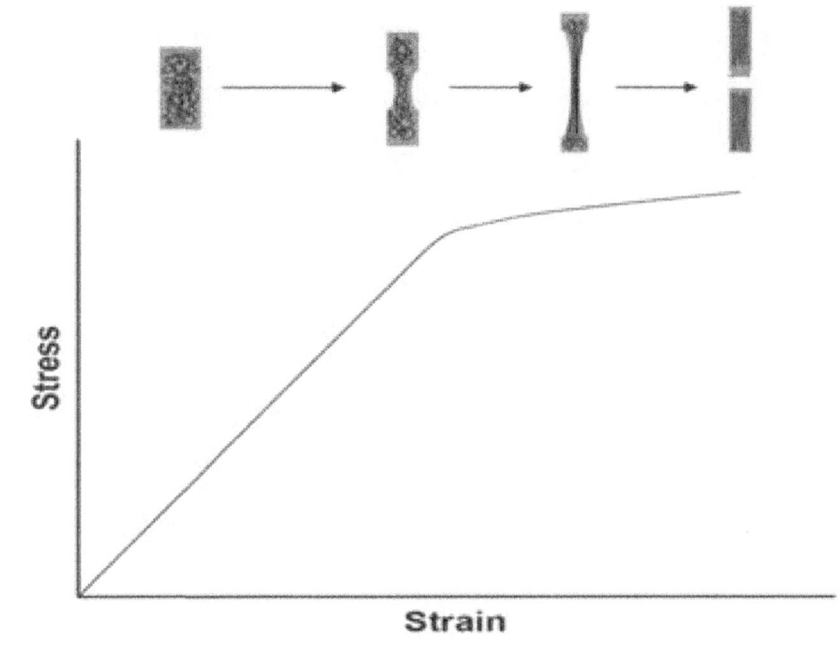

Equation 2

Figure 5-5) by applying Equations 2 and 3 (Felton 2007).

$$\sigma = \frac{F}{A_o}$$

$$\varepsilon = \frac{\Delta L}{L_o}$$ *Equation 3*

Where σ is stress, F is the applied force, A_o is the initial cross-sectional area of the membrane, ε is strain, L_o is the initial length between the grips and ΔL is the increase in the length.

The polymer chains orient themselves in response to the applied stress, transforms through four stages leading to the breakage. Initially, elastic deformation occurs where a linear relationship between the stress and strain occurred. A yield point is then reached. After this point, plastic deformation occurs. (Felton 2007).

Ultimate properties including strain at break are usually the best indicator of material deterioration. Standard mechanical parameters such as young's modulus, extension and force at breaking point, force at breakpoint and yield stress can be measured using a mechanical integrity texture analyser. This technique could indicate the chemical embrittlement of membranes if the values obtained are significantly lower.

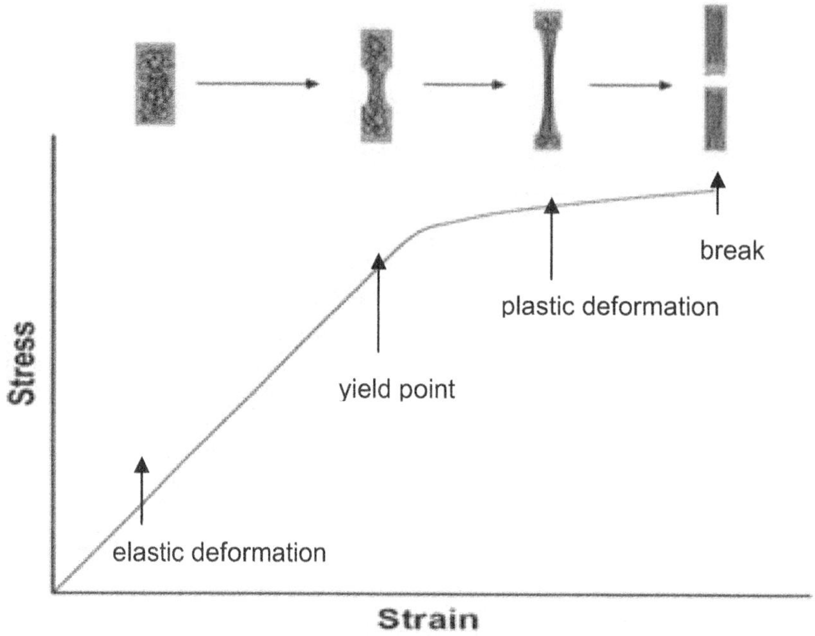

Figure 5-5 Example of a stress-strain curve (Felton 2007).

Past studies on tensile tests have shown deterioration in mechanical strength for PES membranes (MWCO – 20 kDa) decreases with increasing exposure dosage of hypochlorite. The tensile strength at break for the fresh membrane was 16.4 MPa, and decreased to 15.9 MPa and 11.6 MPa for a free chlorine concentration of 1 g.h/L and 50 g.h/L respectively (Arkhangelsky, Kuzmenko et al. 2007).

The embrittlement mechanisms in polymer ageing would be characterised by different behaviours in the stress and yield stress patterns, as shown in Table 5-1 (Gaudichet-Maurin and Thominette 2006).

Table 5-1 Proposed embrittlement mechanisms from tensile tests

(Gaudichet-Maurin and Thominette 2006).

Embrittlement mechanisms	Tensile characteristics
Physical ageing	Increase in yield stress
Crosslinking	Stress increase in the plastic zone
Plasticisation	Decrease in yield stress
Random chain scission	Constancy in yield stress

5.2.1 Surface characterisation techniques

An important direct integrity monitoring technique that has found widespread application in the water treatment industry is the pressure decay test and the porosity tests (Guo, Wyart et al. ; Farahbakhsh 2003). Porosity test could be utilised to observe and measure a wide variety of pore characteristics, while microscopy techniques are widely used in membrane visualisation.

Porosity test

The capillary flow porometer is a powerful and versatile tool to determine pore structure characteristics. It measures the largest and mean flow pore diameters, pore distribution, gas and liquid distribution and gas and liquid permeability. The porometer can be used for both hollow fiber and flat sheet membranes (Jena and Gupta 2002). It is utilised as a direct membrane autopsy technique to observe the effect of chemical agents on the enlargement of pore diameters. The method is also useful to assess whether the effect of chemical agents on pore size and pore size distribution is uniform throughout the membrane.

The principle of the porosity test is to relate the pressure needed to pass liquid or air through the membrane pores. Pressurised non-reacting gas is used to remove liquid from pores and permit gas flow. The differential pressure needed to empty a pore is related to its diameter, D as described in Equation 4.

$$D = \frac{4\,\gamma\cos\theta}{p} \qquad \textit{Equation 4}$$

where γ is the surface tension of wetting liquid, θ is the contact angle (assumed to be close to zero) and p is the differential pressure across the pore (Farahbakhsh 2003). Equation 4 is used along with gas flow rates through wet and dry samples and liquid flow rates through dry samples in measuring other pore characteristics. Figure 5-6 shows the pore characteristics obtainable from measured flow rates and pressures.

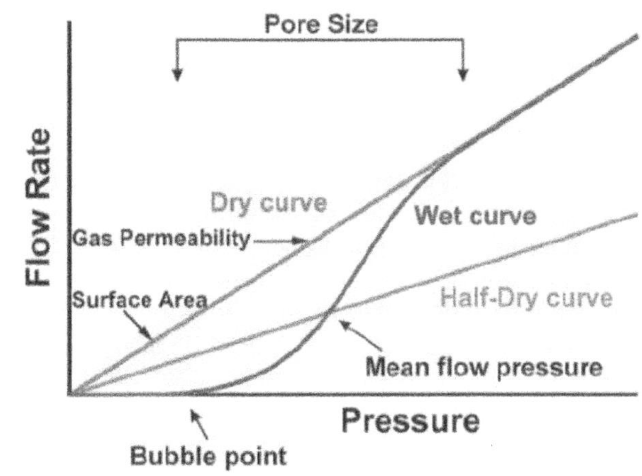

Figure 5-6 Information obtained from the capillary flow porometer

(Jena and Gupta 2002)

Zeta potential analysis

Another technique expected to detect chain scission within the polymer make up is the zeta potential analysis. Zeta potential measurements allow the determination of membrane filtering layer charge sign and charge density (Zeng, Ye et al. 2006).

Zeta potential is measured by applying a controlled electric field across the sample suspension via electrodes. Particles featuring electric charge migrate toward the electrode of opposite polarity. The magnitude of zeta potential is proportional to the velocity of the particles (Malvern Instrument Ltd 2010).

Zhu and Nyström (1998) investigated the effects of exposure of PS membranes to several cleaning agents: NaOH, ammonium hydroxide (NH_4OH), machine powder (MP) consisting of phosphates, silicates and carbonates, and NaOCl. Changes in zeta potentials of virgin membranes after cleaning were observed in all cases. Among all cleaning agents tested, NaOH resulted in the highest reduction while only little change of zeta potentials were observed for NH_4OH indicating that cleaning agents modified PS membranes but to different extent according to membrane types and cleaning agent (Zhu and Nyström 1998).

Contact angle analysis

Contact angle analysis is the measurement of the angle at which a liquid/vapour interface meets a solid surface and could be used to determine the hydrophilicity of a membrane, as seen in Figure 5-7. Differences in contact angle due to the process of "hydrophilization" in manufacturing, where the hydrophilicity of the membrane is modified to suit the application need (Liu, Caothien et al. 2006). For an example, the relatively hydrophobic PVDF membrane can be made more hydrophilic by surface modification through a proprietary oxidation step (Liu, Caothien et al. 2006).

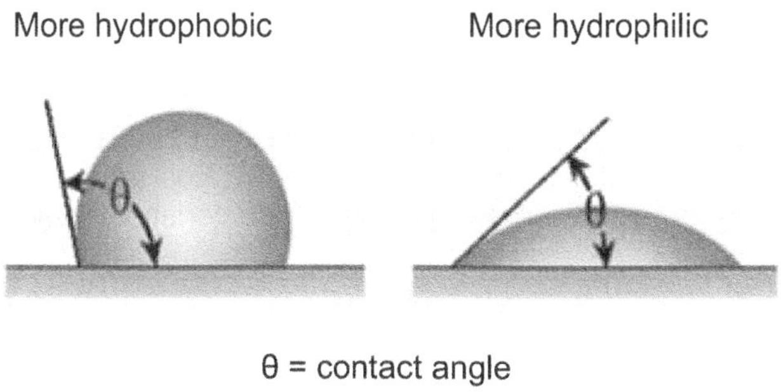

Figure 5-7 Contact angle analysis.

Previous studies have reported that chlorine attach during cleaning could result in greater flux due to increase in hydrophilicity (Kwon and Leckie 2006; Arkhangelsky, Kuzmenko et al. 2007; Simon, Nghiem et al. 2009). For example, the contact angle values of cleaned PES membrane reduced as the hypochlorite concentration increased, indicating that the membrane becomes more hydrophilic. The contact angle before hypochlorite exposure was 58, gradually decreased to 55, 48, 46, 44 and 43 with increasing hypochlorite exposure dosages of 20, 40, 60, 80 and 100 g-h/L. The decrease in contact angle was could be due to (Arkhangelsky, Kuzmenko et al. 2007):

- the partial ionisation of the membrane active layer, forming charged phenyl sulphonate (PhSO3-), more hydrophilic than the uncharged ones and therefore responsible for the increase of hydrophilicity of the membrane.
- the formation of bigger pores due to membrane degradation, making the water drop to spread out by capillary forces.

Contact angle measurements helps in measuring in the hydrophilicity/hydrophobicity of the membrane and the results can be strongly affected by other parameters such as pore size/porosity.

Electron microscopy (EM)

Scanning electros microscopy (SEM) produces very high-resolution images when particle beams of electron illuminate a specimen, revealing details less than 1 to 5 nm in size. From the highly-magnified image generated, detailed membrane characterisation can be performed in terms of its membrane symmetry, diameter, thickness and pore sizes.

Rouaix *et al.* (2006) studied the microscopic morphology changes occurring on PS membrane surface using SEM. The SEM image for the fresh membrane is compared with that of the membrane immersed in 100 ppm hypochlorite solution (for 16 days at 25°C) in Figure 5-8, and the resulting texture of the membrane surface was filamentous and spongy respectively.

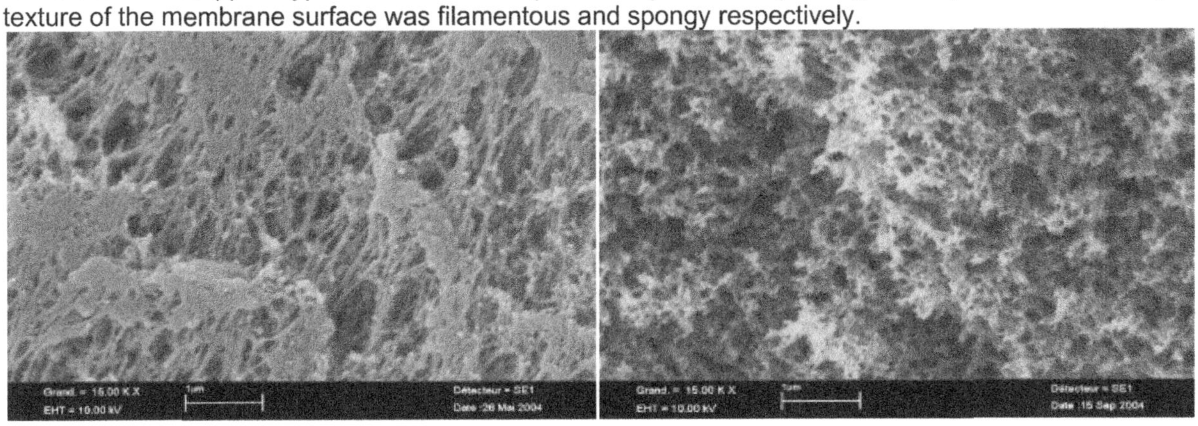

(a) (b)

Figure 5-8 SEM images of (a) fresh membrane and (b) membrane immersed for 16 days at 100 ppm hypochlorite at 25 C

(Rouaix, Causserand et al. 2006).

In addition to SEM, transmission electron microscopy (TEM), scanning tunnelling microscopy (STM), field emission scanning electron microscopy (FESEM), confocal scanning laser microscopy (CSLM) were also reported for membrane characterisation, membrane fouling and degradation (Le-Clech, Chen et al. 2006; Pontié, Thekkedath et al. 2007; Tang, Kwon et al. 2007; Wyart, Georges et al. 2008).

Atomic force microscopy (AFM)

AFM is employed as a powerful tool to study membrane surface morphology. AFM produces micrograph images of the membrane surface and provide useful information about pore size distribution (MF/UF), surface roughness (NF/RO), surface forces, electrical properties and interactions between membrane and foulant (Hilal, Mohammad et al. 2003; Arkhangelsky, Kuzmenko et al. 2007). AFM is a one of the surface probe microscope techniques in which a precision tip interacts with the sample and the forces acting between them is measured. The tip is be attached to the free end of cantilever. When the tip is brought close to the membrane surface, attractive or repulsive forces operate between the precision tip and the membrane surface. The resulting interactive force makes a positive or negative bending of cantilever. Thus, the interactions are measured from the degree of cantilever bending using a laser beam (Vilalta-Clemente and Gloystein 2008). A representative diagram explaining the working of AFM is presented in Figure 5-9.

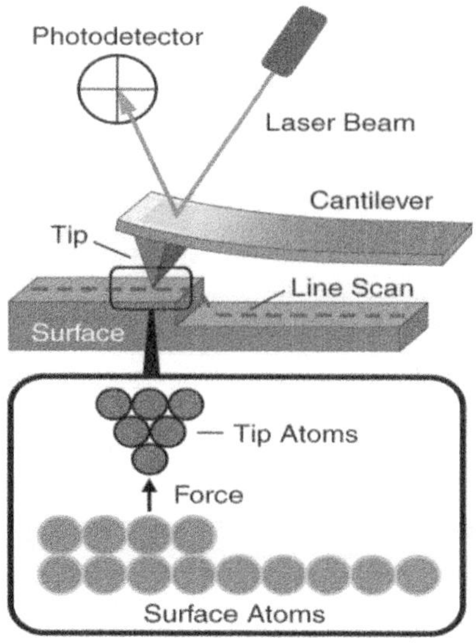

Figure 5-9 Working principle of AFM (Vilalta-Clemente and Gloystein 2008).

Arkhangelsky et al (2007) compared the active surface layers of fresh membranes with chlorine treated PES membrane, shown in Figure 5-10. The surface roughness is calculated as the peak-to-valley distance, which is the distance between the highest data point and the lowest data point, was found to be 14.5 nm for fresh membrane and 9.8 nm for the fouled membrane. As a qualitative measure, size of the sink was related to the membrane pore and the pore size. It is seen from the AFM images that the chlorine damaged surface possess larger sinks, suggesting the larger pore size (Arkhangelsky, Kuzmenko et al. 2007)

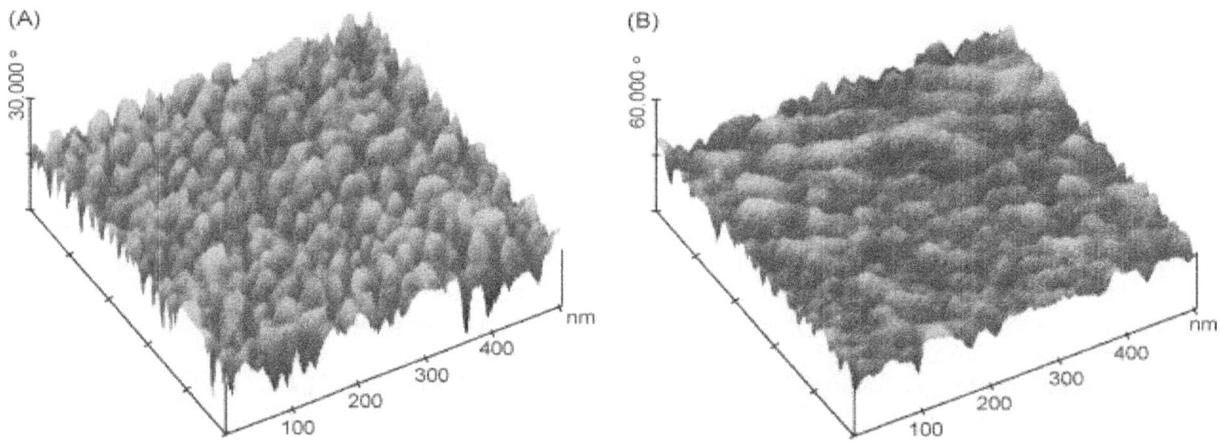

Figure 5-10 AFM images of (a) fresh and (b) chlorine treated PES membrane active layer (5μm x 5μm)

(Arkhangelsky, Kuzmenko et al. 2007).

Small Angle X-Ray Scattering (SAXS)

SAXS is an X-ray collimation system, where elastic scattering of X-rays by the test sample is recorded at very low angles. The angular range is typically 0.1 to 10°, and therefore provides information about shape and size of the sample. SAXS is having the advantage over conventional X-ray diffraction spectroscopy, in terms of its ability to examine the non-crystalline substances. Synchrotron SAXS can be a powerful tool to study the structural changes in mesoscopic and nanoscopic scales during polymer degradation (Chu and Hsiao 2001). Application of SAXS in membrane characterisation is limited, which could be due to the high cost associated with the measurement.

5.2.2 Chemical assessment

In addition to the changes in the physical properties could be associated with chemical changes occurring with the membrane material during membrane degradation. Some of the analytical techniques employed in assessing the chemical modification are discussed in this section.

X-Ray photoelectron spectroscopy (XPS)

XPS is a surface-sensitive technique that measures the elemental composition (except hydrogen) and provides chemical binding information for the top 1–5 nm of the surface region. XPS determines the surface composition of membrane samples and demonstrates their relative changes upon chemical treatment. XPS analysis of virgin and aged membrane allows the detection of chemical agent depositions on membrane surfaces and missing functional groups as a result of chemical reaction between chemical agents and the membrane surface.

The XPS spectra obtained for the fresh, BSA fouled and hypochlorite-cleaned PES membranes is presented in Figure 5-11, where in a nitrogen peak is seen only for the fresh and fouled membrane (Arkhangelsky, Kuzmenko et al. 2007).

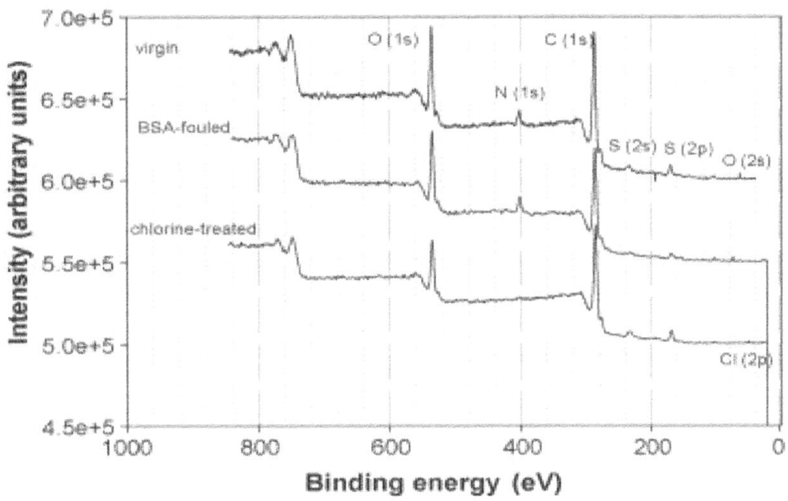

Figure 5-11 XPS image of fresh BSA fouled and hypochlorite cleaned PES membrane

(Arkhangelsky, Kuzmenko et al. 2007).

The presence of the nitrogen peak is probably due to the PVP, an additive used in membrane manufacture, to increase the hydrophilicity. Thus, the absence of nitrogen peak suggests the dislodging of PVP from the membrane during hypochlorite exposure. After membrane cleaning, small amount of chlorine was also detected on to the membrane surface. The elemental composition calculated from XPS indicated that, in response to hypochlorite cleaning, carbon and sulfur concentrations significantly increased and oxygen concentration decreased. (Arkhangelsky, Kuzmenko et al. 2007)

XPS analysis conducted on PVDF aged by NaOCl showed a significant decrease of oxygen (13% to 1%) in 21 weeks, implying the elimination of carbonyl and carboxylic groups (Puspitasari, Granville et al. 2010). It was concluded that these functional groups probably belong to the surface modifying substances expected to increase membrane hydrophilicity.

Attenuated total reflection - Fourier Transform infrared (ATR-FTIR) spectroscopy

ATR-FTIR allows the detection of changes in the functional groups of the membrane by detecting bonds between atoms in a molecule and thus providing information about functional groups and chemical structures of the membrane (Kwon and Leckie 2006). Each peak illustrates a specific type of molecular vibration: symmetric, asymmetric, twisting, wagging, bending or rocking (Garton 1992).

The FTIR spectra is generally analysed with the high resolution Hummel polymer and additives spectrum library available with the FTIR program (Puspitasari, Granville et al. 2010).

FTIR spectra of hydrophobic membranes like PVDF normally shows only carbon and fluorine functional groups (Flösch, Lehmann et al. 1992). Hydrophilic PVDF membranes revealed the presence of organic groups used in the coating substances as surface modifiers (Puspitasari, Granville et al. 2010). Although less quantitative than XPS, ATR-FTIR provides significant qualitative details about the types of functional groups, making it a highly complementary method to XPS (Kwon and Leckie 2006).

ATR-FTIR spectroscopy has been previously applied to assess the mechanism of PES membrane degradation by Arkhangelsky et al (2007). From the spectrum obtained for the fresh and hypochlorite-exposed membranes, formation of phenyl sulphonate and the weakening of C-S bond on hypochlorite exposure was detected through the decrease in the C-S stretching vibration at 1485 cm^{-1}.

Thermal analysis

The thermo-gravimetric analysis (TGA) is used to detect the changes in the mass of the sample as a function of temperature and/or time. TGA is commonly used to determine the degradation temperatures of the polymer materials. Chemical modification in the polymeric structure during ageing are expected to bring a change in the degradation temperature, can be accessed from TGA profiles of the membrane before and after exposure to the ageing chemicals. TGA would generate a plot of weight loss versus temperature (temperature-scan curve), which is related to the composition of the membrane and the thermo-gravimetric property of each component (Tay, Liu et al. 2003). The presence of different materials in a sample would be shown by different temperature loss gradients. The materials present in the sample could be predicted by comparing the temperature and mass loss curve as the material starts to decompose to values obtained by the degradation curve of a pure sample of the expected material. The amount of each material present in the samples was calculated as a function of percentage of mass loss in the particular degradation region per total mass of sample.

Gel permeation chromatography (GPC)

The impact of chemical cleaning on membrane is mostly directed through the polymer backbone modification like, chain breaking and scission reactions. This generally results in the decrease of polymer chain length and therefore the distribution of molecular weight. In those instances, the molar mass distribution can be measured before and after chemical cleaning, using GPC. In this technique, polymers are dissolved in suitable solvent and the elution fractions are analysed by suitable detectors.

Molar mass distribution of PS was determined by Rouaix *et al.*, (2006) before and after exposure to hypochlorite using tetrahydrofuran solvent and the fractions were detected by differential refractometer. It was observed that PS was selectively dissolved from PS/PVP membrane; the chromatograms obtained for the new and hypochlorite exposed membranes are shown in Figure 5-12. With increase in the duration of hypochlorite exposure, the molar mass distribution was decreasing, showing the chain breaking resulting in the formation of low molar mass polymers. For the fresh membrane, the distribution was highest, 78.3 kg mol^{-1} and decreased to 74, 60, 59 and 52 kg mol^{-1} after 11, 32, 64 and 128 days respectively.

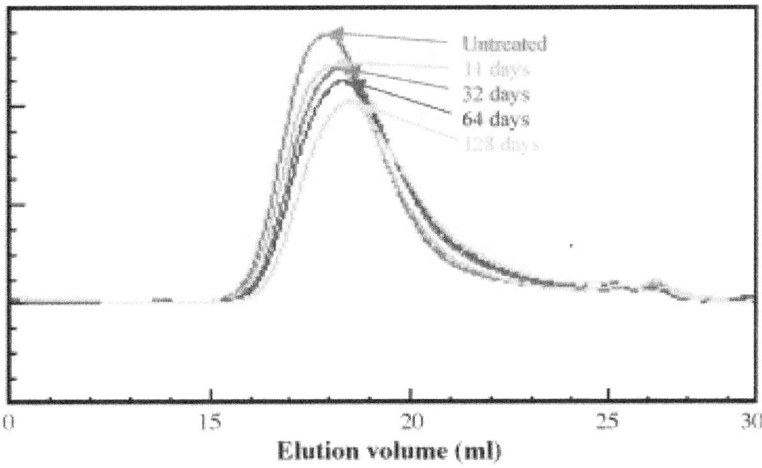

Figure 5-12 Chromatograms of new and hypochlorite exposed membranes, 100 ppm hypochlorite, pH 8 for different durations

(Rouaix, Causserand et al. 2006)

Table 5-2 Summary of analytical techniques for assessing membrane ageing.

Technique	**Information derivable**	**limitations**
Flux recovery and membrane resistance	Changes in intrinsic membrane hydraulic resistance and permeability due to ageing	-
Permeate quality testing	Efficiency of cleaning chemicals	-
Tensile testing	Changes in mechanical strength of the membrane	-
Porosity measurement	Changes in pore structure characteristic of the membrane	-
Zeta potential measurement	Changes in membrane charge and charge density of the membrane surface	-
Contact angle measurement	Changes in hydrophilicity/hydrophobocity of the membrane	-
SEM	Changes in membrane symmetry, diameter, thickness and pore size	-
AFM	Changes in pore size distribution, surface roughness, surface forces, electrical properties and membrane –foulant interactions	Tedious instrumentation
SAXS	Changes in membrane surface structure at mesoscopic and nanoscopic scales	High cost
XPS	Changes in the elemental composition of the membrane surface	High cost
ATR-FTIR	Changes in the chemical structure of membrane	Membrane additives are to be clearly known to study the membrane modification.
TGA	Change in thermal stability of membrane constituents mass as a function of temperature and/or time	The membrane compositions are to be clearly known to compare with standards and understand the thermal behaviour
GPC	Change in molar mass with membrane degradation	Solvent selection is critical as the membrane components and their compositions are proprietary.

6. ASSESSMENT OF RELIABILITY AND SENSITIVITY OF ANALYTICAL TECHNIQUES

The first section of this study aims to assess the reliability of the selected membrane characterisation techniques. Tests were carried out on five new samples of both the 0.22 µm FS/PVDF and 0.04 µm HF/PVDF membranes. The average value of the samples was calculated for each test. The sample standard deviations, s, of each characterisation method were then computed to find the variability in results from the mean value,

$$s = \sqrt{\frac{1}{n-1}\sum_{i=1}^{n}(x_i - x)^2} \qquad \textit{Equation 5}$$

{x1, x2, ..., xn} are the observed values and *x* is the mean value of these observations. The average and standard deviation values of the new membranes are represented as follows:

New membrane results: $\mathbf{A \pm S_A}$

The reliability of the results was measured by calculating the percentage of the standard deviation of the mean values:

$$reliability = \frac{S_A}{A} \times 100\% \qquad \textit{Equation 6}$$

Accelerated membrane ageing experiments were then performed to observe the sensitivity of each technique in detecting changes on the membrane as a result of exposure to the cleaning solutions. The two membranes were immersed in 7% NaOCl solution for 1 week and membrane characterisation was then conducted on the membranes. The average and standard deviation values of the aged membranes were then calculated. The average and standard deviation values of the aged membranes are represented as follows:

Aged membrane results: $B \pm S_B$

The sensitivity of the membrane ageing experiment in this study was analysed through three different approaches:

Approach 1: Percentage change analysis

The most straightforward procedure to assess sensitivity is to calculate the percentage change of the average results of the new (A) and aged (B) samples. The equation is shown as:

$$\text{Percentage change} = \frac{|B - A|}{|A|} \times 100\% \qquad \textit{Equation 7}$$

Approach 2: Percentage change with standard deviations

Approach 1 however does not take into account the standard deviations associated with multiple samples of the new (S_A) and aged (S_B) membranes. A stricter procedure is presented in Approach 2, where the standard deviation (S_{total}) was calculated along with the percentage change

$$S_{total} = \left(\sqrt{\left(\frac{\sqrt{S_B^2 + S_A^2}}{B - A}\right)^2 + \left(\frac{S_A}{A}\right)^2}\right) \times \text{Percentage change} \qquad \textit{Equation 8}$$

Approach 3: T-test analysis

The T-test is a statistical comparison technique which determines whether the means of two groups are statistically different from each other (Douglas C. Montgomery 2004). T value was calculated based on the equation:

$$t = \frac{B - A}{\sqrt{\frac{S_B^{\,2}}{n_B} + \frac{S_A^{\,2}}{n_A}}}$$ *Equation 9*

n = number of samples

The likelihood that the means obtained from the two groups are significantly different could be calculated from the t-value in terms of Confidence Interval (CI). The calculated t-value is then compared to a t-distribution value chart at a chosen level of significance. If the calculated confidence level is equal or greater than the value at the level of significance, the observed means are concluded to be significantly different.

6.1 Clean water test

Information obtained from the experiment includes the varying flux at constant TMP for flat sheet and the TMP values at constant flux for the hollow fibre membranes. The calculation for membrane resistance was based on Equation below:

$$R = \frac{\Delta P}{\eta J}$$ *Equation 10*

R = membrane resistance (m^{-1})

ΔP =Trans membrane pressure (TMP) (Pa)

η = viscosity (Pa.s)

J = flux (ms^{-1})

6.1.1 Results

Table 6.1: Membrane resistance tests of new and aged PVDF membranes.

		Flat sheet	Hollow fibre
Membrane resistance, R_m (m^{-1} x 10^9)	New	21.0 ± 4.0	670 ± 200
	Aged	6.7 ± 2.0	370 ± 30
Reliability (%)	Standard deviations	19	6
Sensitivity (%)	Percentage change	213	63
	Percentage change + standard deviation	213 ± 98	63 ± 9
	T-test (CI)	99	100

From Table 6.1, it was observed that the membrane resistances (R_m) for both flat sheet and hollow fibre PVDF membranes decreased significantly after ageing in sodium hypochlorite solution.

The average error in the reliability analysis was observed to be higher for flat sheet as compared to hollow fibre membranes. The different degrees of reliability between the two types of membranes were expected as the membranes had different specifications and different setups were used in the clean water tests.

For the sensitivity analysis, both flat sheet and hollow fibre membranes showed more than 50% change in membrane resistance after membrane ageing. Confidence interval from t-tests also showed high sensitivity of the clean water test in measuring the change in membrane resistance after ageing. There was more than 99% probability that the mean resistance value of the aged membranes is significantly different from the new membranes.

6.2 Porometry

6.2.1 Results

The nominal pore size of a membrane is defined as the maximum pore size of the virgin membrane. The membranes were rinsed with water and subjected to the clean water test before sampling. Rinsing would result in the pores opening up, and increased pore sizes are expected for the membranes.

The mean pore size diameter and bubble point diameter were automatically obtained from the instrument. Results were not able to be obtained for the 0.04 µm hollow fibre membranes. This is most probably because the pore sizes are too small to be detected by the porometer. The minimum pore size detected by the porometer for the flat sheet membrane was approximately 0.04 µm, as shown in Figure 6-1.

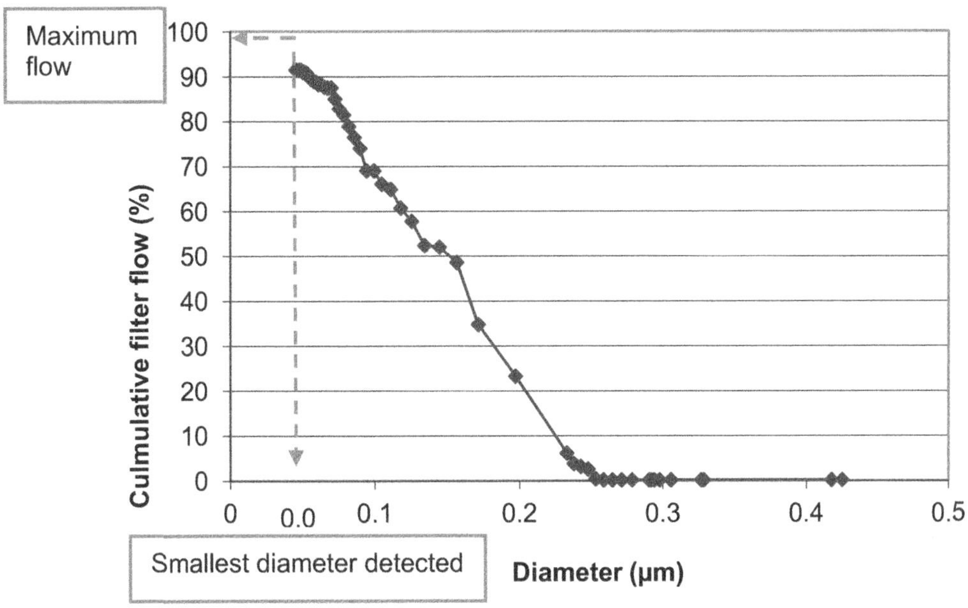

Figure 6-1: Cumulative filter flow graph obtained from porometry analysis.

The flat sheet membrane samples were tested using the dry up / wet up method. The porometry results for flat sheet membranes are shown in Table 6.2.

Table 6.2: Porometry tests of new and aged PVDF membranes.

		Flat sheet	
		Mean pore diameter	Bubble point diameter
Results (µm)	New	0.146 ± 0.020	0.433 ± 0.009
	Aged	0.125 ± 0.035	0.436 ± 0.005
Reliability (%)	Standard deviations	17	2
Sensitivity (%)	Percentage change	14	0.7
	Percentage change + standard deviation	14 ± 38	0.7 ± 3.0
	T-test (CI)	69	38

From reliability analysis, the average error for mean pore diameter was observed to be much higher for mean pore diameter calculations compared to the bubble pore diameter. The high standard deviation for mean pore diameter was probably due to the uneven distribution of membrane pore sizes across the samples.

The mean pore diameter was shown to be more susceptible to change compared to the bubble point diameter. The sensitivity analysis however also showed high standard deviation for both sets of result, leading to low levels of significant difference between the new and aged membranes. The inconsistent mean pore size results could occur as a result of the porometry analysis managing to only detect 90% of the pore diameters, as was shown in Figure 6-1. The remaining 10% were undetected as the sizes were less than approximately 0.04 µm, reducing the accuracy of the method. The observed changes were not statistically significant to make any conclusion on the effect of membrane ageing on the membrane pore sizes.

6.3 Contact angle

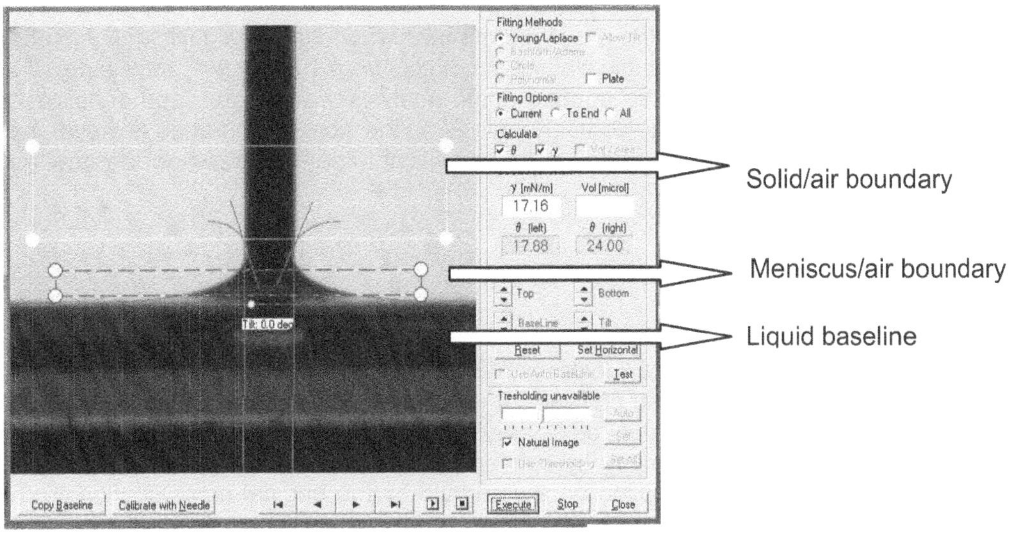

Figure 6-2: Contact angle analysis curve fitting window for hollow fibre membranes.

The recorded contact angle images for hollow fibre membranes needed to be manually adjusted. From the curve fitting window, the liquid baseline (red line), meniscus/air boundary (blue box) and solid/air boundary (yellow box) were fixed. The contact angle values were then calculated by the software, and an average of the left and right meniscus values were calculated.

Preliminary tests showed that the contact angle values for meniscus formed when the solid first touch the liquid surface to be very inconsistent with standard deviation of more than 50%. The contact angle values for meniscus formed as the solid leaves the liquid surface were found to be more reliable and hence this method was used for the hollow fibre experiments.

6.3.1 Results

The values for left and right contact angles were averaged and readings were repeated thrice for each sample. The calculated values for contact angle analysis were presented in Table 6.3.

Table 6.3: Contact angle analysis of new and aged PVDF membranes.

		Flat sheet	Hollow fibre
Contact angle (°)	New	67 ± 2	32 ± 7
	Aged	94 ± 5	35 ± 5
Reliability (%)	Standard deviations	3	20
Sensitivity (%)	Percentage change	40	7
	Percentage change + standard deviation	40 ± 10	7 ± 13
	T-test (CI)	100	51

Good reproducibility was observed for flat sheet membranes, with average standard deviations less than 5 percent. Contact angle measurements for hollow fibre membranes however showed great variability, with an average error of 20 percent. The inconsistencies mainly arise from the manual measurement technique which allowed a great room for error. The method used to collect contact angle values needs to be improved to reduce the variability in readings and enhance reproducibility of the results.

As seen in Table 6.3, the hydrophobicity of flat sheet membranes was found to significantly increase after membrane ageing, with the contact angle values changing from 67° to 94°; indicating a 40 percent change. It can be concluded that the contact angle analysis was reliable and sensitive to membrane ageing for flat sheet samples. The contact angle of hollow fibre membranes however was shown to vary greatly from one sample to another, with insignificant changes between the new and aged membranes.

6.4 Tensile test

Stress-strain curve

The breaking force, a measure of the maximum amount of load that a sample can withstand before it breaks, was obtained directly from the texture analyser. It was however only useful in comparing the strength of samples with the same configuration, cross sectional area and length. The load-elongation curve needs to be normalised for fibre geometry by converting into a stress-strain curve. The stress (σ) can be obtained by dividing the load applied on the sample by the original cross sectional area where it is acting upon:

Equation 11

$$\sigma = F / A_0$$

The engineering strain was derived from measuring the deformation of the sample, in this case the elongation of the membrane. It was calculated by the ratio of change in length ΔL of the material to the initial length L_0:

Equation 12

$$e = \Delta L / L_0 = (l - L_0)/ L_0$$

An example of a stress-strain curve is shown in Figure 6-3. The linear portion of the curve is the elastic region, and the slope is the Young's Modulus or the modulus of elasticity. After the polymer has lost its elastic properties, the material continues to deform until it breaks.

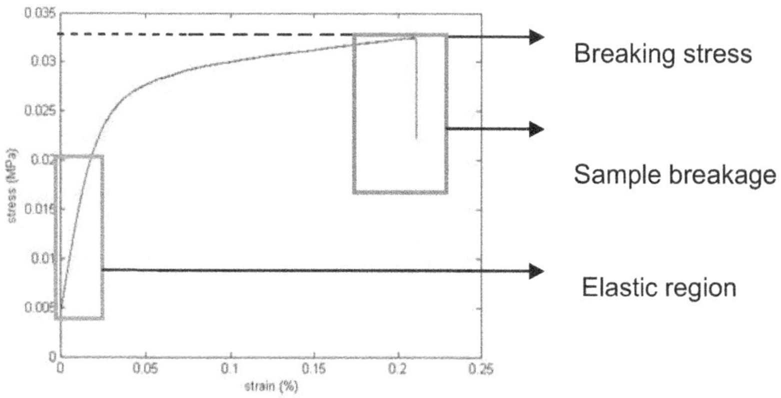

Figure 6-3: Example of a stress-strain curve plotted in Matlab.

Breaking stress

Useful information that could be obtained from a stress strain curve include the breaking stress, maximum elongation and the modulus of elasticity. The tensile strength is the maximum stress given

Equation 13

by the stress strain curve (breaking stress, σ_{max}), which could also be calculated by dividing the breaking force (F_{max}) with the original cross sectional area, A_0

$$\sigma_{max} = F_{max} / A_0$$

Breaking factor

The standard test method for tensile properties of thin plastic sheeting (ASTM D882) suggested the calculation of the breaking factor of a sample instead of breaking stress for very thin films (0.13 mm and less) as the breaking load (F_{max}) and tensile properties may not be proportional to cross-sectional area. Breaking factor was calculated by dividing the maximum load by the original minimum width of the specimen:

$$\text{Breaking factor, } f = F_{max} / \text{width}_{initial} \qquad \textit{Equation 14}$$

The breaking factor was calculated for the flat sheet PVDF samples with nominal thickness of 0.125 mm.

Ultimate elongation

Material elongation (e) is similar to the strain and is given as the ratio of the stretched lenath L to the original length L_0:

$$e = L / L_0 \times 100\% \qquad \textit{Equation 15}$$

The ultimate elongation is the maximum elongation before sample breakage whilst the elastic elongation is the maximum elongation without permanent deformation of the material. In assessing membrane failure, the membrane ultimate elongation was assessed.

Young's Modulus

Young's Modulus (modulus of elasticity) measures the stiffness of an isotropic elastic material. It is defined as the ratio of the uniaxial stress over strain in which the Hooke's Law holds. The modulus could be experimentally determined from the slope of a stress-strain curve created during tensile tests conducted on a sample of the material at the elastic region, as shown in Figure 6-3. The determination of Young's Modulus could be performed by fitting a linear curve on the stress-strain plot either manually or automatically. An automatic procedure was chosen for this study by performing least squared linear regression analysis using Matlab (regstats). A linear curve was fitted from the first point on the stress-strain plot (x_0, y_0) to the maximum point (x_{max}, y_{max}). The process was repeated for (x_0, y_0) to (x_{max-i}, y_{max-i}) until the coefficient of determination (r^2) value is equal or higher than 0.9. The method is reliable given that there is no anomaly in the stress-strain curve, e.g. the presence of more than one linear region in the curve, as the program would not be able to distinguish the elastic region from the others.

An alternative analysis method was to plot the curve on Microsoft Excel or similar programs and to manually select the elastic region and fit a linear curve. The data selection range was adjusted until the fitted linear curve has a value of r^2 equal or higher than 0.9. A disadvantage of this method however is the inconsistency in data selection between different stress-strain curve and the longer time needed to process the data one at a time, whereas the automatic analysis would be able to run all the analyses simultaneously. The r^2 values could be varied between 0.9 and 1 as the calculated modulus values and standard deviation would change depending on the sample.

6.4.1 Results

The breaking force, breaking stress, maximum elongation and Young's Modulus values were calculated for both flat sheet and hollow fibre membranes. An additional parameter, the breaking factor was calculated for the flat sheet samples. The results for the tensile properties before and after ageing were tabulated in Table 6.4 and

Table 6.5.

Table 6.4: Breaking force, breaking factor and breaking stress of new and aged PVDF membranes.

Tensile properties		Breaking force (N)		Breaking factor (N/mm)	Breaking stress (MPa)	
		FS	HF	FS	FS	HF
New		8.7 ± 0.6	2.7 ± 0.1	0.87 ± 0.06	6.7 ± 0.4	5.4 ± 0.2
Aged		8.5 ± 0.1	2.7 ± 0.2	0.85 ± 0.01	6.7 ± 0.1	5.4 ± 0.3
Reliability (%)	Standard deviations	7	3	7	6	4
Sensitivity (%)	Percentage change	3	0.003	3	0.4	0.004
	Percentage change + standard deviation	3 ± 9	0.003 ± 10	3 ± 9	0.4 ± 7.0	0.004 ± 10
	T-test (CI)	51	0	51	0	0

Table 6.5: Maximum elongation and Young's Modulus of new and aged PVDF membranes.

Tensile properties		Max. Elongation (mm)		Young's Modulus (MPa) $r^2 = 0.90$		Young's Modulus (MPa) $r^2 = 0.99$	
		FS	HF	FS	HF	FS	HF
New		136 ± 13	241 ± 6	69 ± 20	30 ± 8	121 ± 40	47 ± 4
Aged		124 ± 6	233 ± 2	76 ± 24	35 ± 6	155 ± 51	41 ± 2
Reliability (%)	Standard deviations	10	10	29	25	33	10
Sensitivity (%)	Percentage change	9	3	11	15	28	7
	Percentage change + standard deviation	9 ± 14	3 ± 4	11 ± 65	15 ± 44	28 ± 78	7 ± 13
	T-test (CI)	78	90	33	64	66	84

From Table 6.4 and

Table 6.5, high reliability was observed for breaking force, breaking stress and ultimate elongation for both flat sheet and hollow fibre membranes. For the sensitivity analysis, it was observed that the biggest percentage change occurred in the calculations of Young's Modulus for flat sheet membranes, with more than 10% change.

For hollow fibre membranes, the percentage change varied from 7 to 15% depending on r^2 values. T-tests however showed that maximum elongation of the membranes as the most sensitive tensile characterisation technique. The results were due to the high degree of variance from the Young's Modulus calculations. The breaking force and breaking stress showed some changes after ageing for flat sheet membranes, 3 and 0.4% respectively. No significant change was observed for hollow fibre membranes, as confirmed by the t-test.

6.5 Thermogravimetric analysis

TGA allows the determination of thermal stability of materials and their compositions by measuring the amount and rate of mass change of a sample as a function of temperature or time.

Compositional analysis of samples was performed by observing the thermal degradation curve obtained from the test. The presence of different materials in a sample would be shown by different temperature loss gradients. The materials present in the sample could be predicted by comparing the temperature and mass loss curve as the material starts to decompose to values obtained by the degradation curve of a pure sample of the expected material. The amount of each material present in the samples was calculated as a function of percentage of mass loss in the particular degradation region per total mass of sample.

Comparison of thermal stability between samples of the same material compositions was carried out by plotting the thermal degradation curves together and observing the trend of material degradation. Mass loss for samples of lower thermal stability would occur earlier than the other samples. The temperature at which the maximum rate of material decomposition occur was observed as the peak of the derivative mass loss as a function of temperature or time.

6.5.1 Results

Figure 6-4 showed the plot of percentage membrane mass loss (Y1) and the rate of mass loss (Y2) as a function of temperature for a new flat sheet PVDF membrane. From the plot, it was observed that flat sheet PVDF membranes exhibited a one-step degradation mechanism which indicates that only one material is present in the membrane. The mass loss due to decomposition of the material started at 498°C, with 77% mass loss occurring between 498°C to 550 °C. The maximum rate of decomposition occurred at 525 °C.

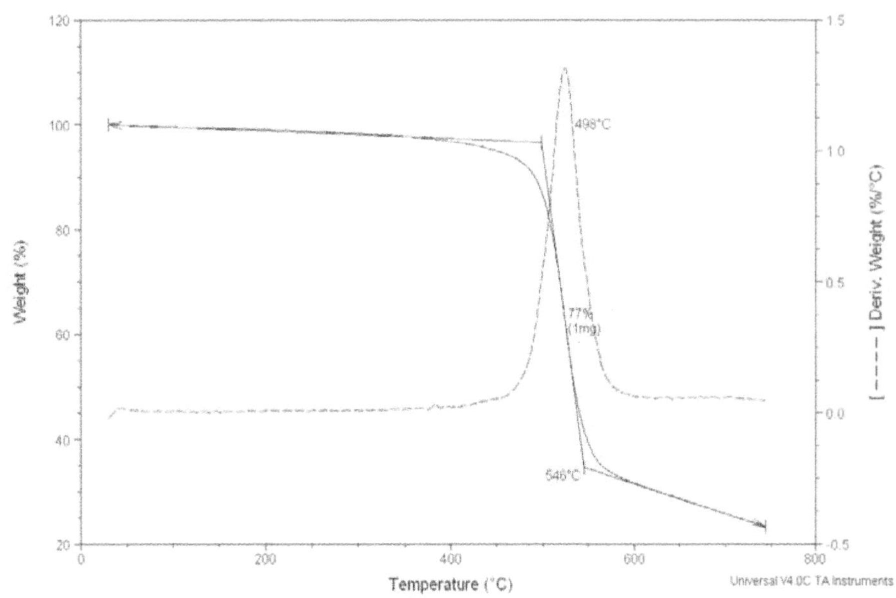

Figure 6-4: TGA results of new flat sheet PVDF membranes.

The plot of percentage membrane mass loss (Y1) and the rate of mass loss (Y2) as a function of temperature for a new hollow fibre PVDF membrane were shown in Figure 6-5.

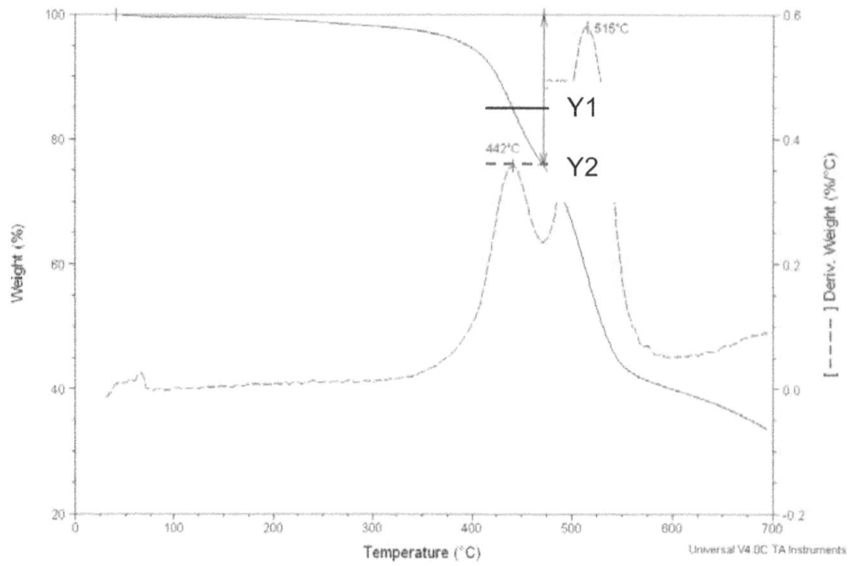

Figure 6-5: TGA results of new hollow fibre PVDF membranes (Y1: percentage sample mass loss, Y2: rate of mass loss

Hollow fibre PVDF membranes exhibited a two-step degradation mechanism as observed from the two peaks in the rate of mass loss (Y2) curve. It indicated the presence of at least two materials in the sample. Decomposition of the first material occurred between 388-460°C, with 24% mass loss of the sample. The maximum rate of decomposition occurred at 442°C. The decomposition of the second material started at 460°C and continued to degrade until 700°C, with the maximum rate of decomposition at 515 °C.

The thermal stability results of new and aged PVDF membranes were compared in terms of temperature with maximum degradation rate and percentage mass loss of the samples. Temperatures with 5% and 10% weight loss were also included for a more detailed analysis of the materials in the sample. The literature values and experimental results for PVDF membranes were summarised in Table 6.6 ((Sina Ebnesajjad 2005))

Table 6.6: Thermal properties of new and aged PVDF membranes.

Thermal properties		Temp. with maximum degradation rate (°C)		Percentage mass loss (%)	
		FS	HF	FS	HF
Literature values (Sina Ebnesajjad 2005)		491		85	
New		528 ± 4	432 ± 9 519 ± 4	75 ± 2	22 ± 2 48 ± 2
Aged		528 ± 13	461 ± 3 537 ± 6	74 ± 1	12 ± 2 67 ± 2
Reliability (%)	Standard deviations	0.6	2 0.7	3	9
Sensitivity (%)	Percentage change	0	7 1	1	45
	Percentage change + standard deviation	0 ± 0	7 ± 3 1 ± 3	1 ± 4	45 ± 19
	T-test (CI)	0	99	43	99

The literature values of TGA of pure PVDF polymers in N_2 atmosphere and the experimental results were compared. It was reported that 82% weight loss of pure PVDF occurred between 399 to 508°C, with the maximum degradation rate occurring at 491°C (Sina Ebnesajjad 2005). The flat sheet membranes exhibited similar thermal degradation properties except for the temperature with maximum degradation rate. The experimental values were 8% higher than the expected temperature. The difference could be attributed to additives and polymeric blends added to the membrane which would result in a shift to the degradation curve.

The hollow fibre membranes were found to have a two-step degradation mechanism. The first part of the curve showed a degradation temperature lower than the expected values for PVDF polymers. The average temperatures with five percent weight loss for the hollow fibre membranes were 392°C and 408°C for new and aged membranes respectively; less than 50°C from the literature value of pure PVDF. It was postulated that this material is an additive to the membrane. The most common additives added to increase membrane hydrophilicity are PVP and PEG. The thermo-gravimetric properties of these additives in nitrogen atmosphere were summarised in Table 6.7. It was observed that the average temperature with maximum degradation rate for the first peak of new hollow fibre membranes at 432°C has a less than 3% difference with both PVP and PEG. Further analysis would be needed to confirm which material is indeed the additive used.

Table 6.7: TGA properties of common membrane additives in N_2 atmosphere.

	Temperature degradation range (°C)	Temperature with maximum degradation rate (°C)	Ref.
PVP	280-550	420	(Peniche, Zaldívar et al. 1993)
PEG	340-445	428	(Wang, Ma et al. 2001)

The reliability of TGA was measured by comparing the temperature with maximum degradation rate for the samples. As a two-step degradation mechanism was observed for hollow fibre membranes, the values for both peaks were analysed. High reproducibility was observed for both membranes, with average standard deviation less than 5%. A visual comparison on the mass loss against temperature curves showed very similar degradation trends across the five samples tested for each type of membrane.

In analysing the sensitivity of the thermal properties of new and aged membrane samples, the mass loss degradation curves were plotted together and the compared. The material degradation trend was observed and analysed. The plot for hollow fibre membranes was shown in Figure 6-6.

From Figure 6-6, it was observed that thermal degradation for aged membranes started later than the new membranes, with the temperature of first maximum degradation rate at 464°C and 442°C respectively. The percentage mass loss for the first material was reduced from 24 to 12%. The temperature at which the second maximum degradation rate occurred also increased from 515 to 532 °C. Increasing temperature of maximum decomposition rate could be attributed to changes occurring at the molecular level. Increased rigidity in the polymeric chains after hypochlorite exposure could have shifted the thermal stability curve to the right.

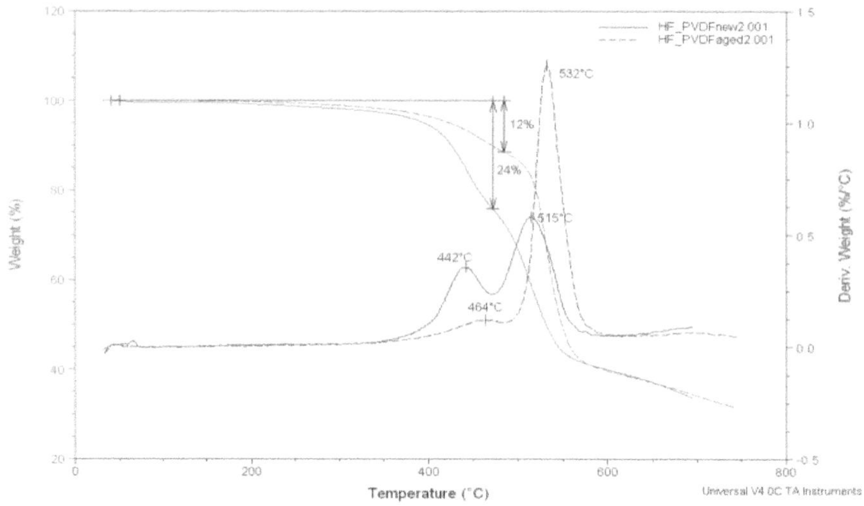

Figure 6-6: Thermal properties of new and aged hollow fibre PVDF membranes.

As seen in Table 6.6, the percentage changes for flat sheet membrane were less than 1% for all thermal stability aspects studied. TGA was not sensitive to the changes on the flat sheet membrane thermal properties. More significant changes were observed for hollow fibre membranes, with 8% shift in the temperature range for material decomposition and 7% increase for the temperature with maximum degradation rate. The greatest percentage change was observed in the percentage mass loss of the first material detected, with a 45% difference between the new and aged PVDF membranes.

7. COMPARISON OF THE MEMBRANE CHARACTERISATION TECHNIQUES

The objective of this section is to compare the suitability of the different membrane characterisation techniques in detecting membrane degradation. The comparison was performed by developing a ranking system, with each criterion being given a weighting factor depending on its importance.

The main aspects considered were the reliability and sensitivity of the membrane characterisation techniques as discussed in part 4.1. Reliability comparison was carried out by comparing the average and maximum errors of the methods, as summarised in Table 7.1.

Table 7.1: Summary of the reliability assessment on new PVDF membrane.

<table>
<tr><th colspan="2" rowspan="2"></th><th colspan="2">Average standard deviations (%)</th></tr>
<tr><th>Flat sheet</th><th>Hollow fibre</th></tr>
<tr><td colspan="2">Clean water test</td><td>19</td><td>6</td></tr>
<tr><td rowspan="2">Porometry</td><td>Mean pore diameter</td><td>17</td><td rowspan="2">-</td></tr>
<tr><td>Bubble pore diameter</td><td>2</td></tr>
<tr><td colspan="2">Contact angle</td><td>3</td><td>20</td></tr>
<tr><td rowspan="5">Tensile test</td><td>Breaking force</td><td>7</td><td>3</td></tr>
<tr><td>Breaking stress</td><td>6</td><td>4</td></tr>
<tr><td>Ultimate elongation</td><td>10</td><td>10</td></tr>
<tr><td>Young's Modulus
(r^2 = 0.90)</td><td>29</td><td>25</td></tr>
<tr><td>Young's Modulus
(r^2 = 0.99)</td><td>33</td><td>10</td></tr>
<tr><td colspan="2">TGA*</td><td>3</td><td>9</td></tr>
</table>

* Percentage mass loss (%)

The sensitivity analysis evaluated whether the properties of the new and aged samples are significantly different using three approaches: percentage change, percentage change with standard deviation and the t-test. The summary of the assessment was presented in Table 7.2.

Table 7.2: Summary of the sensitivity assessment on new and aged PVDF membranes.

<table>
<tr><th colspan="2" rowspan="2"></th><th colspan="3">Flat sheet</th><th colspan="3">Hollow fibre</th></tr>
<tr><th>Percentage change (%)</th><th>% change + st. dev. (%)</th><th>CI from t-test (%)</th><th>Percentage change (%)</th><th>% change + st. dev. (%)</th><th>CI from t-test (%)</th></tr>
<tr><td colspan="2">Clean water test</td><td>213</td><td>213 ± 98</td><td>99</td><td>63</td><td>63 ± 9</td><td>100</td></tr>
<tr><td rowspan="2">Porometry</td><td>Mean pore diameter</td><td>14</td><td>14 ± 38</td><td>69</td><td colspan="3" rowspan="2">-</td></tr>
<tr><td>Bubble pore diameter</td><td>0.7</td><td>0.7 ± 3</td><td>38</td></tr>
<tr><td colspan="2">Contact angle</td><td>40</td><td>40 ± 3</td><td>100</td><td>7</td><td>7 ± 13</td><td>51</td></tr>
<tr><td rowspan="5">Tensile test</td><td>Breaking force</td><td>3</td><td>3 ± 9</td><td>51</td><td>0.003</td><td>0.003 ± 10</td><td>0</td></tr>
<tr><td>Breaking stress</td><td>0.4</td><td>0.4 ± 7</td><td>0</td><td>0.004</td><td>0.004 ± 10</td><td>0</td></tr>
<tr><td>Ultimate elongation</td><td>9</td><td>9 ± 14</td><td>78</td><td>3</td><td>3 ± 4</td><td>90</td></tr>
<tr><td>Young's Modulus (r^2 = 0.90)</td><td>11</td><td>11 ± 65</td><td>33</td><td>15</td><td>15 ± 44</td><td>64</td></tr>
<tr><td>Young's Modulus (r^2 = 0.99)</td><td>28</td><td>28 ± 78</td><td>66</td><td>7</td><td>7 ± 13</td><td>84</td></tr>
<tr><td colspan="2">TGA*</td><td>1</td><td>1 ± 4</td><td>43</td><td>45</td><td>45 ± 19</td><td>99</td></tr>
</table>

* Percentage mass loss (%)

An analysis was carried out using points system to compare the sensitivity of each membrane characterisation technique using the different approaches. The point guideline was given in Table 7.3. Three points (***) were given to indicate high sensitivity, two points for moderate and one point (*) for low. Zero point indicated that the method is considered not sensitive to membrane ageing. The sensitivity analysis was summarised in

Table 7.4

From

Table 7.4 it was observed that approach 2 is not a viable method to perform the comparison as it was too stringent in analysing the sensitivity of the membrane characterisation techniques. An advantage of approach 3 (t-test) over approach 1 (percentage change) is that it takes into account the variances and the number of samples of both groups of data obtained. Approach 3 was therefore chosen as the most suitable sensitivity analysis technique for both the flat sheet and hollow fibre membranes.

Table 7.3: Points guideline for sensitivity analysis using different approaches.

Approach	1	2	3
Points	Percentage change (%)	% change + standard deviation	t-test (CI, %)
***	> 15	% change: >15 and % change > standard deviation	CI > 90%
**	10-15	% change: 10-15 and % change > standard deviation	CI: 70%-90%
*	5-9	% change: 5-9 and % change > standard deviation	CI: 50%-69%
-	< 5	% change: <5 or % change < standard deviation	CI < 50%

Table 7.4: Results summary of sensitivity analysis on new and aged PVDF membranes.

		Flat sheet			Hollow fibre		
Approach		1	2	3	1	2	3
		% change	% change + st. dev	t-test	% change	% change + st. dev	t-test
Clean water test		***	***	***	***	***	***
Porometry	Mean pore diameter	**	-	*	n/a		
	Bubble pore diameter	-	-	-			
Contact angle		***	***	***	*	-	*
Tensile test	Breaking force	-	-	*	-	-	-
	Breaking stress	-	-	-	-	-	-
	Ultimate elongation	*	-	**	-	-	***
	Young's Modulus (r^2 = 0.90)	**	-	-	**	-	*
	Young's Modulus (r^2= 0.99)	***	-	*	*	-	**
TGA		-	-	-	***	***	***

Other criteria taken into account include the time needed to prepare the samples, average test duration and result processing. Tests with samples that required minimal preparation and shorter testing time are generally more preferable. Some of the membrane characterisation instruments presented the required result automatically, while others necessitate further data processing or interpretation. Automatic results are superior to manual data processing and interpretation, as human error can be eliminated and the subjectivity of data interpretation removed. The summary of these criteria is presented in Table 7.5.

Table 7.5: Summary of sample preparation, average test duration and results processing considered in ranking membrane characterisation techniques.

<table>
<tr><th colspan="2"></th><th>Sample preparation</th><th>Av. test duration (min)</th><th>Results processing</th></tr>
<tr><td colspan="2">Clean water test</td><td>Potting hollow fibre membranes: required 24 hours to completely dry</td><td>5</td><td>Manual calculation of R_m based on flux and TMP values obtained.</td></tr>
<tr><td colspan="2">Porometry</td><td>Dry samples</td><td>50</td><td>Automatic.</td></tr>
<tr><td colspan="2">Contact angle</td><td>Requires changes in instrument configuration & calibration before testing HF membranes</td><td>5</td><td>Results obtained after manual adjustment of baseline and phase boundaries. Possible random error.</td></tr>
<tr><td rowspan="4">Tensile test</td><td>Breaking force</td><td rowspan="4">Wet samples: requires 24 hours to soak in Mili-Q</td><td rowspan="4">5</td><td>Automatic.</td></tr>
<tr><td>Breaking stress</td><td rowspan="2">Results obtained after processing data with Matlab/spreadsheet programs.</td></tr>
<tr><td>Ultimate elongation</td></tr>
<tr><td>Young's Modulus</td><td>Results obtained after processing data with Matlab/spreadsheet programs. Possibility of error from curve fitting.</td></tr>
<tr><td colspan="2">TGA</td><td>Dry samples</td><td>50</td><td>Automatic values from software, but needed interpretations.</td></tr>
</table>

7.1 Ranking of the techniques

A weighting system was developed by dividing the criteria into high (3 points), moderate (2 points) and low importance (1 point) in choosing the most suitable characterisation method. The points were given in terms of number of stars for each aspect. The guideline for the number of points was summarised in Table 7.6.

Table 7.6: Points guideline for each aspect considered in technique comparison.

Stars	Reliability (Av. Error %)	Sensitivity (t-test,CI %)	Sample preparation (time)	Test duration (time)	Results processing
****	< 5%				
***	5-10%	CI > 90%	<15 min	<5 min	Automatic results from test
**	11-25%	CI: 70-90%	15-60 min	5-30 min	Some result processing needed
*	> 25%	CI: 50-69%	>60 min	>30 min	Result processing and interpretation needed

The ranking summaries for flat sheet and hollow fibre membranes were given in Table 7.7 and

Table 7.8 respectively. The points for each method were obtained by multiplying the importance points with the number of stars given. For an example, the reliability of clean water test for flat sheet membranes was given 2x2 = 4 points, and the sensitivity was given 3x3 = 9 points. The total points for each membrane characterisation technique were calculated by adding the points of each aspect under consideration. From the total points, the most suitable membrane characterisation methods were recommended.

Table 7.7: Ranking of membrane characterisation techniques for flat sheet membranes.

<table>
<tr><th colspan="2"></th><th>Reliability</th><th>Sensitivity</th><th>Sample preparation</th><th>Test duration</th><th>Results processing</th><th>Total *</th><th>Recommended</th></tr>
<tr><td colspan="2">Importance (3 =high, 2 = moderate, 1 = low)</td><td>2</td><td>3</td><td>1</td><td>1</td><td>1</td><td></td><td></td></tr>
<tr><td colspan="2">Clean water test</td><td>**</td><td>***</td><td>***</td><td>***</td><td>**</td><td>21</td><td>✓✓</td></tr>
<tr><td rowspan="2">Porometry</td><td>Mean pore diameter</td><td>**</td><td>*</td><td rowspan="2">***</td><td rowspan="2">*</td><td>***</td><td>14</td><td></td></tr>
<tr><td>Bubble pore diameter</td><td>****</td><td></td><td>***</td><td>15</td><td></td></tr>
<tr><td colspan="2">Contact angle</td><td>****</td><td>***</td><td>***</td><td>***</td><td>**</td><td>25</td><td>✓✓</td></tr>
<tr><td rowspan="5">Tensile test</td><td>Breaking force</td><td>***</td><td>*</td><td rowspan="5">*</td><td rowspan="5">***</td><td>***</td><td>16</td><td>✓</td></tr>
<tr><td>Breaking stress</td><td>***</td><td></td><td>**</td><td>12</td><td></td></tr>
<tr><td>Ultimate elongation</td><td>***</td><td>**</td><td>**</td><td>18</td><td>✓</td></tr>
<tr><td>Young's Modulus (r^2 = 0.90)</td><td>*</td><td></td><td>*</td><td>7</td><td></td></tr>
<tr><td>Young's Modulus (r^2= 0.99)</td><td>*</td><td>*</td><td>*</td><td>10</td><td></td></tr>
<tr><td colspan="2">TGA</td><td>****</td><td></td><td>***</td><td>*</td><td>*</td><td>13</td><td></td></tr>
</table>

From Table 7.7, it was concluded that the most suitable technique to assess membrane ageing for flat sheet membrane is contact angle analysis, followed by clean water test and the calculations of breaking force and ultimate elongation from tensile test.

Porometry analysis was not recommended as it gave a high standard deviation for mean pore size diameter in the reliability and sensitivity tests. The issue could be due to pore size variations in different samples. For the tensile properties, the breaking stress remained mostly constant after membrane ageing and Young's Modulus encountered high variance between results. TGA was found to be reliable; however it did not pick up any significant change to the membrane thermal properties after the accelerated ageing process.

Table 7.8: Ranking of membrane characterisation techniques for hollow fibre membranes.

<table>
<tr><th colspan="2"></th><th>Reliability</th><th>Sensitivity</th><th>Sample preparation</th><th>Test duration</th><th>Data processing</th><th>Total *</th><th>Recommended</th></tr>
<tr><td colspan="2">Importance (3 =high, 2 = moderate, 1 = low)</td><td>2</td><td>3</td><td>1</td><td>1</td><td>1</td><td></td><td></td></tr>
<tr><td colspan="2">Clean water test</td><td>***</td><td>***</td><td>*</td><td>***</td><td>**</td><td>21</td><td>✓✓</td></tr>
<tr><td colspan="2">Contact angle</td><td>**</td><td>*</td><td>**</td><td>***</td><td>*</td><td>13</td><td></td></tr>
<tr><td rowspan="6">Tensile test</td><td>Breaking force</td><td>****</td><td></td><td rowspan="6">*</td><td rowspan="6">***</td><td>***</td><td>15</td><td></td></tr>
<tr><td>Breaking stress</td><td>****</td><td></td><td>**</td><td>14</td><td></td></tr>
<tr><td>Ultimate elongation</td><td>***</td><td>***</td><td>**</td><td>21</td><td>✓✓</td></tr>
<tr><td>Young's Modulus (r^2 = 0.90)</td><td>*</td><td>*</td><td>*</td><td>10</td><td></td></tr>
<tr><td>Young's Modulus (r^2 = 0.99)</td><td>***</td><td>**</td><td>*</td><td>17</td><td>✓</td></tr>
<tr><td colspan="2">TGA</td><td>***</td><td>***</td><td>***</td><td>*</td><td>*</td><td>20</td><td>✓✓</td></tr>
</table>

Table 7.8 presented the ranking results for the tested hollow fibre membranes. Clean water test, measurement of ultimate elongation from tensile test and TGA were recommended as they have high degree of reliability and sensitivity as well as ease of sampling and measurement. Young's Modulus required high degree of data processing and interpretations but could be used to supplement the results from other techniques.

Other tensile properties such as breaking force and breaking stress are not recommended as they showed low degrees of sensitivity from the tests. Porometry analysis was not possible due to the small pore sizes of the membranes tested. Contact angle analysis showed poor reliability and sensitivity, mostly due to the high standard deviation between results. Refinement in the measurement techniques could be carried out in future work to achieve more significant outcomes.

8. CONCLUSIONS

Experimental reliability of the membrane characterisation techniques was rigorously assessed for the flat sheet and hollow fibre membranes. It was found that the suitability of different membrane characterisation techniques as predictive tools for membrane ageing depends on membrane configuration (flat sheet/hollow fibre) and type (material, average pore size).

Clean water test was found to be more reliable for hollow fibre membranes, while contact angle was more suited for flat sheet membranes. The standard deviations for tensile tests and TGA were found to be mostly similar between both types of membrane. Porometry analysis was only possible for the flat sheet membranes as the pore size of the hollow fibre membranes were too small to be detected by the instrument. The results of the reliability assessment are also affected by data interpretation. Young's Modulus was calculated based on a curve-fitting analysis which introduced further errors in the results obtained directly from the measurements.

The chemical, morphological and mechanical changes to the membrane might not be to the same extent after the ageing process. It would be reflected by the different degrees of sensitivity of the characterisation techniques in detecting the differences between new and aged membranes.

The sensitivity of the techniques also depended on membrane configurations and materials. T-tests were performed on the PVDF samples to assess the sensitivity of each technique in detecting membrane degradation. The most sensitive techniques for the flat sheet membranes were the clean water test and contact angle analysis. Mean pore diameter, breaking force and ultimate elongation were also suitable in complementing the results. For hollow fibre membranes, the most significant changes on membrane properties were observed through the clean water test, TGA and measurement of ultimate elongation from the tensile test. Contact angle and Young's Modulus results were less sensitive to the degradation process.

Polymers have different levels of tolerance towards chemical exposure as reflected by the ageing results. PVDF membranes are generally described as chlorine resistant, while PP membranes have moderate tolerance towards hypochlorite. The accelerated ageing tests in the study resulted in the degradation of both types of membrane.

A thorough evaluation of the membrane characterisation techniques was executed by developing a ranking system to compare different aspects in selecting the most suitable method to assess membrane ageing. The summary of the results was presented in Table 8.1.

Table 8.1: Recommended characterisation techniques for membrane ageing analysis.

Techniques		Flat sheet	Hollow fibre
Clean water test		✓✓	✓✓
Porometry	Mean pore diameter		
	Bubble pore diameter		
Contact angle		✓✓	
Tensile test	Breaking force	✓	
	Breaking stress		
	Ultimate elongation	✓	✓✓
	Young's Modulus (r^2 = 0.90)		
	Young's Modulus (r^2 = 0.99)		✓
TGA			✓✓

9. LIMITATIONS AND RECOMMENDATIONS

Clean water tests, contact angle, porometry, tensile and TGA were completed on both flat sheet and hollow fibre membranes. The different material configurations would result in different degrees of reliability and suitability in the membrane ageing assessment. The inconsistency limits the comparison between the same materials of different configurations.

Some of the techniques like contact angle and porometry analysis were more suitable for the flat sheet membranes. There is currently no existing procedure for a reliable contact angle analysis of hollow fibre membranes using the available technology in the laboratories. A methodology needs to be developed to increase the reliability of the technique. Tensile properties of flat sheet membranes were uncommonly measured using the texture analyser. The results of different configurations were not comparable for this technique.

Only PVDF membranes and sodium hypochlorite solutions were used in developing the reliability and sensitivity assessment in detecting membrane ageing. Different polymeric materials have different degrees of sensitivity towards the chemicals. As PVDF is one of the most chemical resistant polymers used in membrane fabrication, some techniques which were not sensitive to the nascent changes on the membrane might be suitable for other materials. Varying cleaning agents would also have different ageing mechanisms, which would affect the suitability of the technique chosen.

In the study, sampling was performed after 7 days for the PVDF membranes and after 2 and 7 days for the PP membranes. It was observed that most of the surface modification substances were already removed for PVDF membranes and the PP membranes have lost most of their mechanical and thermal stability at this point. More frequent sampling could be carried out in future work as critical changes occurring to the membranes might have been missed. Ageing tests at lower concentrations could also allow a better assessment of the nascent changes occurring to the membrane after the exposure to cleaning agents.

A more complete assessment could be performed by comparing assessing a wider range of materials and chemical agents. A wider variety of simple and inexpensive methods could be used to characterise membrane degradation. The methods could also be used in conjunction with some of the novel and emerging technologies as it would allow a better understanding of membrane ageing mechanisms. A summary of recommended characterisation techniques that could be performed in future work is presented in Table 9.1.

Table 9.1: Recommended membrane characterisation techniques for future work.

Techniques	Objectives	Ref.
Gel permeation chromatography (GPC)	Determines the relative molecular weight and distribution of membrane samples.	(Liu, Xu et al. 2009)
Bubble point test / Pressure decay test	Estimates membrane pore sizes and bubble point pressures.	(Farahbakhsh 2003)
X-ray photoelectron spectroscopy (XPS)	Measures the elemental composition, empirical formula, chemical and electronic state of the elements that exist within membrane samples.	(Wolff and Zydney 2004; Kwon and Leckie 2006)
Scanning electron microscopy (SEM)	Produces electronically-magnified image membrane samples for detailed observation.	(Puspitasari, Granville et al.)
Zeta potential analysis	Measures the surface charge and zeta potential of charged particles in membrane samples.	(Kwon and Leckie 2006; Simon, Nghiem et al. 2009)
Binary gas integrity test	Detects the presence of defective or open pores.	(Giglia and Krishnan 2008)

10. REFERENCES

(2008). BCC Research, Report Code: MST047B. Membrane and Separation Technology, Membrane Bioreactors: Global Markets

Alspach, B., S. Adham, et al. (2008). "Microfiltration and ultrafiltration membranes for drinking water." Journal American Water Works Association **100**(12): 84-97.

Amjad, Z. (1993). Reverse osmosis : membrane technology, water chemistry & industrial applications.

Anil K. Pabby, S. S. H. R., Ana Maria Sastre (2009). "Handbook of membrane separations : chemical, pharmaceutical, food, and biotechnological applications."

Argüello, M. A., S. Álvarez, et al. (2003). "Enzymatic cleaning of inorganic ultrafiltration membranes used for whey protein fractionation." Journal of Membrane Science **216**(1-2): 121-134.

Arkhangelsky, E., D. Kuzmenko, et al. (2007). "Hypochlorite Cleaning Causes Degradation of Polymer Membranes " Tribology Letters **28**(2): 109-116.

Arkhangelsky, E., D. Kuzmenko, et al. (2007). "Impact of chemical cleaning on properties and functioning of polyethersulfone membranes." Journal of Membrane Science **305**(1-2): 176-184.

Arnal, J. M., B. Garcia-Fayos, et al. (2009). "Ultrafiltration membrane cleaning with different chemical solutions after treating surface water." Desalination and Water Treatment **7**(July 2009): 198-205.

Asahi Kasei Chemicals. (2006). "Microza - Highly Efficient, Precise Filtration, Durable Module." Retrieved 20th September, 2009, from http://www.asahi-kasei.co.jp/membrane/microza/en/about_microza.html.

ASTM International "ASTM D882 - 09 Standard Test Method for Tensile Properties of Thin Plastic Sheeting."

Baah, C. A. and J. I. Baah (2002). "Polypropylene degradation in NaOH environments." Materials & Design **23**(3): 341-343.

Bansal, B., R. Al-Ali, et al. (2006). "Rinsing and cleaning of [alpha]-lactalbumin fouled MF membranes." Separation and Purification Technology **48**(2): 202-207.

Causserand, C., S. Rouaix, et al. (2008). "Ageing of polysulfone membranes in contact with bleach solution: Role of radical oxidation and of some dissolved metal ions." Chemical Engineering and Processing: Process Intensification **47**(1): 48-56.

Cheryan, M. (1998). Ultrafiltration and microfiltration handbook. Lancaster, Pa., Technomic Pub. Co.

Childress, A. E., P. Le-Clech, et al. (2005). "Mechanical analysis of hollow fiber membrane integrity in water reuse applications." Desalination **180**(1-3): 5-14.

Chu, B. and B. S. Hsiao (2001). "Small-Angle X-ray Scattering of Polymers." Chemical Reviews **101**(6): 1727-1762.

Cox, M. F. (1994). Detergents and Cleaners: A Handbook for Formulators. Munich, Hanser Publishers.

D'Souza, N. M. and A. J. Mawson (2005). "Membrane Cleaning in the Dairy Industry: A Review." Critical Reviews in Food Science and Nutrition **45**: 125-134.

Douglas C. Montgomery, G. C. R., Norma F. Hubele (2004). Engineering Statistics, 4th Edition, Wiley.

Euromba (2006). Membrane bioreactor technology (MBR) with an EU perspective for advanced municipal wastewater treatment strategies for the 21st century Sixth Framework Programme.

Farahbakhsh, K. (2003). "Monitoring the integrity of low-pressure monitoring." Journal American Water Works Association **95**(6): 95-107.

Felton, L. (2007). "Characterisation of coating systems." AAPS PharmSciTech **8**(4): 258-266.

Flösch, D., H. D. Lehmann, et al. (1992). "Surface analysis of poly(vinylidene difluoride) membranes." Journal of Membrane Science **70**(1): 53-63.

Fukatsu, K. and S. Kokot (2001). "Degradation of poly(ethylene oxide) by electro-generated active species in aqueous halide medium." Polymer Degradation and Stability **72**(2): 353-359.

Garton, A. (1992). Infrared Spectroscopy of Polymer Blends, Composites and Surfaces, Hanser.

Gaudichet-Maurin, E. and F. Thominette (2006). "Ageing of polysulfone ultrafiltration membranes in contact with bleach solutions." Journal of Membrane Science **282**(1-2): 198-204.

Giglia, S. and M. Krishnan (2008). "High sensitivity binary gas integrity test for membrane filters." Journal of Membrane Science **323**(1): 60-66.

Gijsbertsen-Abrahamse, A. J., E. R. Cornelissen, et al. (2006). "Fiber failure frequency and causes of hollow fiber integrity loss." Desalination **194**(1-3): 251-258.

Gitis, V., R. C. Haught, et al. (2006). "Application of nanoscale probes for the evaluation of the integrity of ultrafiltration membranes." Journal of Membrane Science **276**(1-2): 185-192.

Guo, H., Y. Wyart, et al. "Low-pressure membrane integrity tests for drinking water treatment: A review." Water Research **In Press, Accepted Manuscript**.

Hashimoto, T. (2005). Application of cylindrical-type hollow fiber module for MBR system, Asahi Kasei Chemicals.

Hilal, N., A. W. Mohammad, et al. (2003). "Using atomic force microscopy towards improvement in nanofiltration membranes properties for desalination pre-treatment: A review." Desalination **157**(1-3): 137-144.

Hong, S. and M. Elimelech (1997). "Chemical and physical aspects of natural organic matter (NOM) fouling of nanofiltration membranes." Journal of Membrane Science **132**(2): 159-181.

IUPAC. (2009). "IUPAC Gold Book - Compendium of Chemical Terminology." Retrieved 16 October, 2009, from http://goldbook.iupac.org/index.html.

Jacangelo, J. G. and R. K.Noack (2005). System concepts. Microfiltlration and ultrafiltration membranes for drinking water. Colorado, American Water Works Association: 51-64.

Jena, A. and K. Gupta (2002) "Flow Porometry: What Can Flow Porometry Do For Us?".

Johnson, J. (2009). "Membrane Fundamentals, Part3: Effect of pH, Temperature Chemicals and Cleaning Procedures vs. Foulant Removal." Retrieved 1st August, 2009, from http://www.waterworld.com/index/display/article-display/276175/s-articles/s-industrial-waterworld/s-volume-7/s-issue-5/s-features/s-membrane-fundamentals-part3-effect-of-ph-temperature-chemicals-and-cleaning-procedures-vs-foulant-removal.html.

Judd, S. (2008). "The status of membrane bioreactor technology." Trends in Biotechnology **26**(2): 109-116.

Judd, S. and C. Judd (2006). The MBR book : principles and applications of membrane bioreactors in water and wastewater treatment. Amsterdam, Elsevier.

Kazemimoghadam, M. and T. Mohammadi (2007). "Chemical cleaning of ultrafiltration membranes in the milk industry." Desalination **204**: 213-218.

Kennedy, M. D., J. Kamanyi, et al. (2008). Water Treatment by Microfiltration and Ultrafiltration. Advanced Membrane Technology and Applications. N. N. Li, A. G. Fane, W. S. W. Ho and T. Matsuura. New Jersey, John Wiley and Sons.

Kimura, K., Y. Hane, et al. "Irreversible membrane fouling during ultrafiltration of surface water." Water Research **38**(14-15): 3431-3441.

Kimura, K., Y. Hane, et al. (2004). "Irreversible membrane fouling during ultrafiltration of surface water." Water Research **38**(14-15): 3431-3441.

Krack, R. (1995). "Chemical Agents and costs in cleaning and disinfection of membrane equipment."

Kraume, M., D. Wedi, et al. (2009). "Fouling in MBR: What use are lab investigations for full scale operation?" Desalination **236**(1-3): 94-103.

Kubota, N., T. Hasimoto, et al. (2008). Microfiltration and Ultrafiltration. Advanced Membrane Technology and Applications. N. N. Li, A. G. Fane, W. S. W. Ho and T. Matsuura. New Jersey, John Wiley & Sons.

Kuzmenko, D., E. Arkhangelsky, et al. (2005). "Chemical cleaning of UF membranes fouled by BSA." Desalination **179**(1-3): 323-333.

Kwon, Y.-N. and J. O. Leckie (2006). "Hypochlorite degradation of crosslinked polyamide membranes: I. Changes in chemical/morphological properties." Journal of Membrane Science **283**(1-2): 21-26.

Le-Clech, P., V. Chen, et al. (2006). "Fouling in membrane bioreactors used in wastewater treatment." Journal of Membrane Science **284**(1-2): 17-53.

Le-Clech, P., A. Fane, et al. (2005). "MBR focus: the operators' perspective." Filtration & Separation **42**(5): 20-23.

Lenntech. (2009). "Membrane cleaning methods." Retrieved November 26, 2009, from http://www.lenntech.com/membrane-cleaning.htm.

Li, X., J. Li, et al. (2005). "Chemical cleaning of PS ultrafilters fouled by the fermentation broth of glutamic acid." Separation and Purification Technology **42**(2): 181-187.

Lim, A. L. and R. Bai (2003). "Membrane fouling and cleaning in microfiltration of activated sludge wastewater." Journal of Membrane Science **216**(1-2): 279-290.

Liu, C., S. Caothien, et al. (2006). "Membrane Chemical Cleaning: From Art to Science." Retrieved 5th July 2009, from http://www.pall.com/pdf/mtcpaper.pdf.

Liu, F., Y.-Y. Xu, et al. (2009). "Preparation of hydrophilic and fouling resistant poly(vinylidene fluoride) hollow fiber membranes." Journal of Membrane Science **345**(1-2): 331-339.

Maartens, A., P. Swart, et al. (1998). "Humic membrane foulants in natural brown water: characterisation and removal." Desalination **115**(3): 215-227.

Madaeni, S. S. and Y. Mansourpanah (2004). "Chemical cleaning of reverse osmosis membranes fouled by whey." Desalination **161**(1): 13-24.

Madaeni, S. S., T. Mohamamdi, et al. (2001). "Chemical cleaning of reverse osmosis membranes." Deslaination **134**(1-3): 77-82.

Malvern Instrument Ltd. (2010). "Zeta potential measurement using laser Doppler electrophoresis (LDE) " Retrieved 4th March, 2010, from http://www.malvern.de/LabGer/technology/zeta_potential/zeta_potential_LDE.htm.

Mohammadi, T., S. S. Madaeni, et al. (2003). "Investigation of membrane fouling." Desalination **153**(1-3): 155-160.

Mulder, M. (1996). Basic Principles of Membrane Technology. The Netherlands, Springer.

New Logic Research. (2007). "Membrane Cleaning 101." Retrieved 6th August 2009, from http://membranecleaner.com.

Oreski, G. and G. M. Wallner (2005). "Aging mechanisms of polymeric films for PV encapsulation." Solar Energy **79**(6): 612-617.

Pearce, G. (2007). "Introduction to membranes: Membrane selection." Filtration & Separation **44**(3): 35-37.

Pearce, G. (2008). "Introduction to membranes-MBRs: Manufacturers' comparison: part 1." Filtration & Separation **45**(2): 28-31.

Pearce, G. (2008). "Introduction to membranes -- MBRs: Manufacturers' comparison: part 2 -- supplier review." Filtration & Separation **45**(3): 30-32.

Pearce, G. (2008). "Introduction to membranes - MBRs: Manufacturers' comparison: part 3 - supplier review." Filtration & Separation **45**(4): 23-25.

Peniche, C., D. Zaldívar, et al. (1993). "Study of the thermal degradation of poly(<I>N</I>-vinyl-2-pyrrolidone) by thermogravimetry-FTIR." Journal of Applied Polymer Science **50**(3): 485-493.

Pilutti, M. and J. E. Nemeth (2006). Technical and cost review of commercially available MF/UF membrane products. International Desalination Association.

Pontié, M., A. Thekkedath, et al. (2007). "Membrane autopsy as a sustainable management of fouling phenomena occurring in MF, UF and NF processes." Desalination **204**(1-3): 155-169.

Puspitasari, V., A. Granville, et al. "Optimum Cleaning and Ageing Effect of Sodium Hypochlorite on Polyvinylidene Fluoride (PVDF) Membrane."

Puspitasari, V., A. Granville, et al. (2010). "Optimum Cleaning and Ageing Effect of Sodium Hypochlorite on Polyvinylidene Fluoride (PVDF) Membrane." Separation and Purification Technology **72**(3): 301-308.

Roesink, H. D. W. (1989). Microfiltration: Membrane Development and Module Design. Enschede, The Netherlands, University of Twente. **PhD**.

Rouaix, S., C. Causserand, et al. (2006). "Experimental study of the effects of hypochlorite on polysulfone membrane properties." Journal of Membrane Science **277**(1-2): 137-147.

Sayed Razavi, S. K., J. L. Harris, et al. (1996). "Fouling and cleaning of membranes in the ultrafiltration of the aqueous extract of soy flour." Journal of Membrane Science **114**(1): 93-104.

Shorrock, C. J. and M. R. Bird (1998). "Membrane Cleaning: Chemically Enhanced Removal of Deposits Formed During Yeast Cell Harvesting." Food and Bioproducts Processing **76**(1): 30-38.

Simon, A., L. D. Nghiem, et al. (2009). "Effects of membrane degradation on the removal of pharmaceutically active compounds (PhACs) by NF/RO filtration processes." Journal of Membrane Science **340**(1-2): 16-25.

Sina Ebnesajjad, P. R. K. (2005). Fluoropolymer Applications in Chemical Processing Industries: The Definitive User's Guide and Databook, William Andrew Publishing/Plastics Design Library.

Tang, C. Y., Y.-N. Kwon, et al. (2007). "Probing the nano- and micro-scales of reverse osmosis membranes--A comprehensive characterisation of physiochemical properties of uncoated and coated membranes by XPS, TEM, ATR-FTIR, and streaming potential measurements." Journal of Membrane Science **287**(1): 146-156.

Tay, J.-H., J. Liu, et al. (2003). "Quantification of membrane fouling using thermogravimetric method." Journal of Membrane Science **217**(1-2): 17-28.

Thominette, F., O. Farnault, et al. (2006). "Ageing of polyethersulfone ultrafiltration membranes in hypochlorite treatment." Desalination **200**(1-3): 7-8.

Trägårdh, G. (1989). "Membrane Cleaning." Desalination **71**(3): 325-335.

Vilalta-Clemente, A. and K. Gloystein. (2008). "Principle of Atomic Force Microscopy (AFM)." Retrieved 7, December, 2009, from http://www.mansic.eu/documents/PAM1/Frangis.pdf.

Wang, F.-Y., C.-C. M. Ma, et al. (2001). "Kinetic parameters of thermal degradation of polyethylene glycol-toughened novolac-type phenolic resin." Journal of Applied Polymer Science **80**(2): 188-196.

Wienk, I. M., E. E. B. Meuleman, et al. (1995). "Chemical treatment of membranes of a polymer blend: Mechanism of the reaction of hypochlorite with poly(vinyl pyrrolidone)." Journal of Polymer Science Part A: Polymer Chemistry **33**(1): 49-54.

Wolff, S. H. and A. L. Zydney (2004). "Effect of bleach on the transport characteristics of polysulfone hemodialyzers." Journal of Membrane Science **243**(1-2): 389-399.

Wyart, Y., G. Georges, et al. (2008). "Membrane characterisation by microscopic methods: Multiscale structure." Journal of Membrane Science **315**(1-2): 82-92.

Yang, Q., T.-S. Chung, et al. (2009). "Microscopic behavior of polyvinylpyrrolidone hydrophilizing agents on phase inversion polyethersulfone hollow fiber membranes for hemofiltration." Journal of Membrane Science **326**(2): 322-331.

Zeng, H. X., H. Q. Ye, et al. (2006). "Characterisation of a hollow-fiber ultrafiltration membrane and control of cleaning procedures by a streaming potential method." Desalination **195**(1-3): 226-234.

Zhang, G. and Z. Liu (2003). "Membrane fouling and cleaning in ultrafiltration of wastewater from banknote printing works." Journal of Membrane Science **211**(2): 235-249.

Zhu, H. and M. Nyström (1998). "Cleaning results characterized by flux, streaming potential and FTIR measurements." Colloids and Surfaces A: Physicochemical and Engineering Aspects **138**(2-3): 309-321.

11. ANNEXURE -1 METHODS AND MATERIALS

This section details the experimental materials and methods used in the study. The experimental set-ups of membrane preparation and ageing with sodium hypochlorite (NaOCl) were explained. Membrane characterisation was performed on new and aged membranes by conducting clean water tests, porometry and contact angle analysis, tensile tests and thermogravimetric analysis (TGA). The procedure of each technique is discussed.

Materials

Membrane

The study utilised microporous flat sheet PVDF membranes and hollow fibre PVDF and PP membranes. The flat sheet used was GVWP Durapore® membrane, a hydrophilic PVDF from Millipore. The membrane thickness is 125 µm with nominal pore size of 0.22 µm. The PVDF hollow fibre membrane employed was 0.04 µm hydrophilic membranes from Memcor, with outer diameter of 0.8 mm and inner diameter of 0.5 mm. The flat sheet membranes were cut into circles of 5 cm diameter to cover the membrane effective area of the dead-end filtration cell.

New hollow fibre PP membranes with pore diameter of 0.2 µm were sourced from Memcor, The outer and inner diameters are 0.65 and 0.39 mm respectively. Fouled PP membranes were also used in the study to compare the hypochlorite ageing effect on new and used membranes. Hydrophilic PP membranes that have been used in the industry for 10 years were obtained from the Sydney Olympic Park Authority. The pore diameter is 0.2 µm and the outer and inner diameters are 0.6 and 0.3 mm respectively. The inorganic compounds found on the membrane surface include aluminium silicates, magnesium, manganese, iron and calcium. The hollow fibre membranes were potted into a loop with both ends of the fibres glued to soft plastic tubing using Araldite®. Potted membranes were air dried for at least 24 hours before being used.

Chemical cleaning agent

Sodium hypochlorite (NaOCl) was used as the chemical cleaning agent in the accelerated cleaning tests. In the study, commercial NaOCl was sourced from Ajax Finechem Pty Ltd with nominal concentration of 12.5% w/v. As NaOCl degrades with time and exposure to light, iodometric titration was performed prior to the ageing process to determine the exact amount of free chlorine in the solution. Dilutions were performed with Mili-Q water to adjust NaOCl concentration to needed values. In hypochlorite solutions, the dominant form of chlorine in water depends on the pH, with HClO dominating at lower pH and increasing ClO^- ions as the solution becomes more alkaline (Rouaix, Causserand et al. 2006).

$$HClO + H_2O \leftrightarrow ClO^- + H_3O^+ \qquad \textit{Equation 16}$$

Experimental setups

Membrane wetting and rinsing

Membrane wetting for hollow fibre membranes was performed with ethanol for 1 min at 200 $Lm^{-2}h^{-1}$, before soaking the membrane in the solution for another 30 mins. The wetted membrane was then rinsed with Mili-Q water extensively. Filtration with Mili-Q water was carried out for 15 mins at 200 $Lm^{-2}h^{-1}$ without permeate recycling. The total organic carbon (TOC) of permeate was measured and compared to Mili-Q water value to ensure remaining ethanol or preservatives were removed. Rinsing and filtration process were repeated if necessary.

Membrane ageing with NaOCl

In the membrane ageing study, the membranes were soaked in NaOCl bath at constant pH and different concentrations. The flat sheet and hollow fibre PVDF membranes were immersed in 7000 ppm NaOCl with pH 12 for 1 week, while the hollow fibre PP membranes were aged in 1000 and 7000 ppm solutions. Samples were taken after 7 days for both PVDF and PP membranes and after 48 hr for the PP membranes. The solution was changed every alternate day to maintain constant concentration. The ageing baths were kept away from natural light during the experiment. All membranes were thoroughly rinsed with Mili-Q water after the ageing process to remove any hypochlorite present before membrane characterisations was carried out.

Membrane characterisation

Following the ageing process, different membrane characterisation techniques were adopted to investigate the suitability of the each method in assessing membrane degradation. Different types of direct and indirect tests were selected; investigating changes in membrane permeability, physical, morphological and chemical properties of the membranes after exposure to cleaning solutions.

Clean water test

Clean water test was performed after the wetting and rinsing procedure of the filtration systems to measure the intrinsic resistance (R_m) of the membrane.

For flat sheet membranes, filtration with 1.5L of Mili-Q water was carried out to obtain pure water flux, J_0. The filtration was operated at a constant trans-membrane pressure (TMP) of 0.5 bar and data was collected every second. Information recorded by the software include the pressure, weight of permeate collected and the calculation for flux at fixed sampling times.

The clean water test for the submerged hollow fibre membranes was performed using the flux stepping method. TMP was calculated as the difference in pressure between the filtrate and the permeate side of the membrane at atmospheric pressure. Constant flux filtration was carried out at 20 $Lm^{-2}h^{-1}$, and increased to 50 and 100 $Lm^{-2}h^{-1}$ after 30-60 s intervals. The pressure was recorded for each flux step, as shown in Figure - A 1.

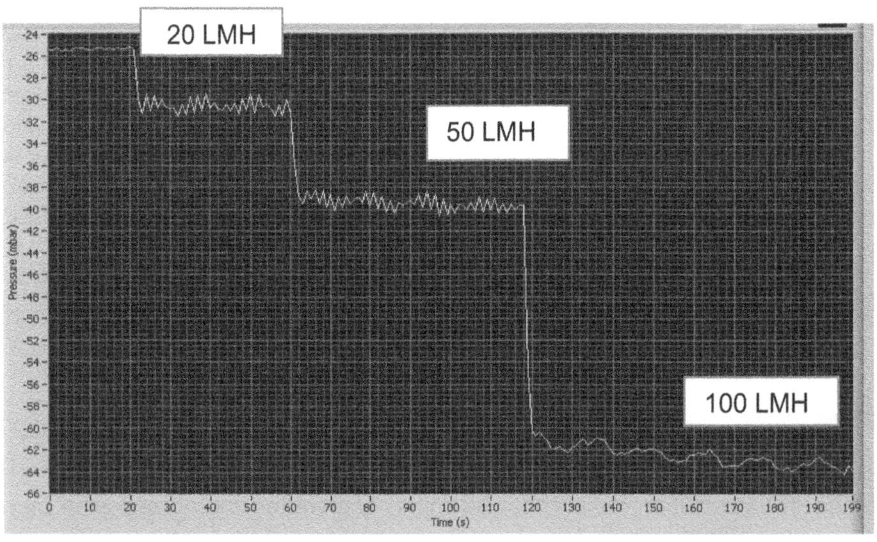

Figure - A 1: Screen shot of the suction pressure values at different flux steps for submerged hollow fibre membrane.

Porometry analysis

Flow porometry is a technique to characterize membrane pore structure and provides information on the average and maximum pore size of the membrane (bubble point diameter). The instrument utilised was the Capillary Flow Porometer 1100 AEL (Porous Material, Inc.) which could be used for a wide variety of applications including filtration media, textiles, powders and membranes (Jena and

Gupta 2002). The instrument is specified to measure pore size between of 0.013 – 500 µm and pressure range within 0 to 500 psi.

The membranes need to be completely dry before the measurements. The samples were cut in order to fit and lie flat in the sample chambers, with the size being bigger than the O-ring to completely cover and ensure proper sealing. For the flat sheet membrane, the samples were cut into a round shape with diameter of 12.7 mm to fit in the sample chamber. The length of sampling for hollow fibre membrane was 5 cm. Only two of the five sample holders were used, and the rest were sealed with Araldite. The fibres were cut longer than the diameter of the sample chamber, glued, and were allowed to dry for 24 h.

The testing procedure chosen for the analysis was the dry up/wet up test. Dry samples were placed in the sample chamber; with plastic adapter plate utilised for flat sheet membranes and a metal plate for hollow fibre membranes. The membrane sample was first introduced to the instrument in dry conditions and a non-reacting gas was passed through the membrane with increasing pressure. The flow rate of air through the pores was measured as a function of differential pressure. The sample was then fully saturated with Galwick™ solution, a non-reactive wetting agent with low surface tension (15.6 dynes/cm) filling the pores and displacing the gas inside the pores. Gas pressure was then utilised to expel the liquid from the pores, and the gas flow rates through the wet samples were measured.

The instrument automatically performed the test execution, data acquisition, storage and reduction. Differential pressure was used to compute the restricted pore diameters of through pores. A curve of the gas flow rate through the dry and wet samples are plotted as a function of pressure.

Contact angle analysis

Contact angle analysis is the measurement of the angle at which a liquid/vapour interface meets a solid surface and could be used to determine the hydrophilicity of a membrane. The contact angle measurement of the new and aged membranes was performed with the KSV 200 CAM goniometer, with Milli-Q water was used as the reference solvent. The goniometer utilised a sample stage, a syringe to apply a droplet of liquid (for flat sheet membranes), a light source to illuminate the droplet, and a set of optics for magnifying the image for observation.

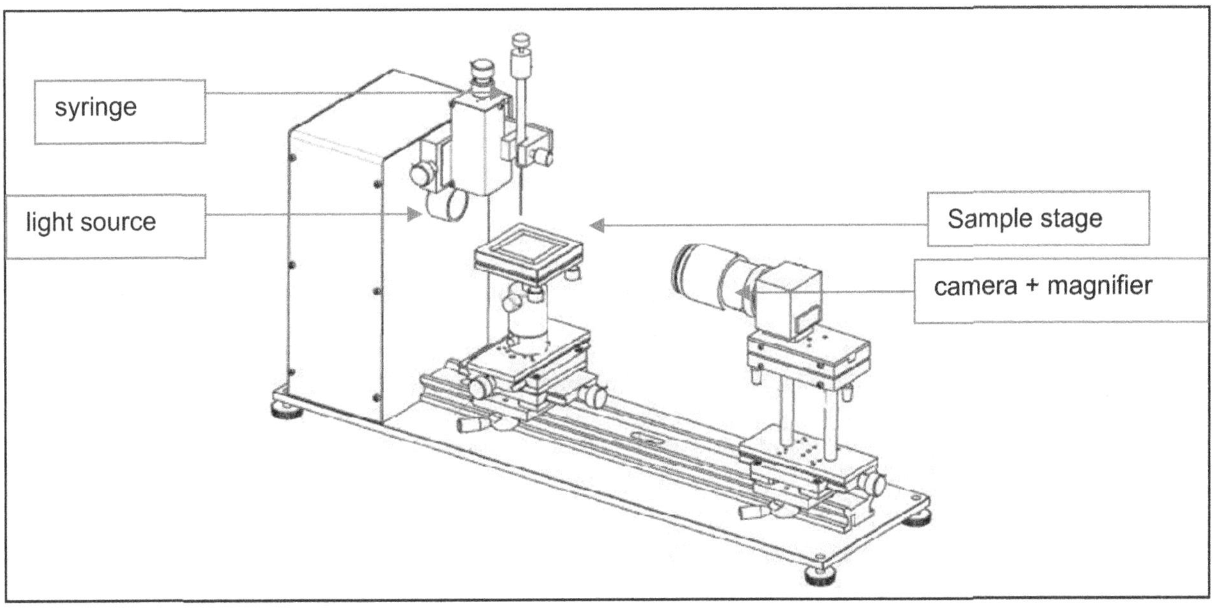

Figure - A 2: Contact angle instrument used in the study (KSV 200 CAM Goniometer).

There are various methods that can be used to measure the contact angle of a sample depending on the sample configuration and surface roughness. The technique used for the flat sheet membrane in the study was the sessile drop method, while meniscus measurement technique was adopted for the hollow fibre membranes.

The sessile drop method is the most common technique used to analyse contact angles on a smooth, flat surface. The samples were air dried before being cut into small pieces and mounted on a flat glass placed on the sample stage. A drop of Mili-Q water was deposited from the syringe pointed vertically down on to the sample surface. A high resolution camera captured the image immediately after the drop was released using the "trigger" function. Contact angle values on both sides of the droplet were automatically measured by the KSV Contact Angle Measurement System software equipped with the instrument. At least five droplets were applied onto duplicate membrane samples to ensure consistency in the readings.

The suggested procedure of measuring contact angle of cylindrical shapes such as rods and fibres with the KSV CAM 200 Goniometer is the meniscus measurement method. The technique is a newly developed method to measure contact angles directly by goniometry as opposed to the more traditional approach of using force tensiometry. The force tensiometric method measures the interaction forces present when a sample of solid is brought into contact with the test liquid and calculates the contact angle from the forces of interaction, solid geometry and surface tension of the liquid (Yang, Chung et al. 2009).

The meniscus method involved replacing the syringe with the investigated solid at the syringe clamp and placing water on the sample stage. Sample was lowered until it intersects with the water surface and a meniscus is formed. Using a trigger function, images were recorded once the meniscus is visible. After two seconds, the solid was taken out and image recording was stopped once the meniscus disappeared.

In the preliminary stages of the experiment, the solid was moved manually and resulted in high levels of human error, as the speed and degree of the solid touching the water surface vary greatly and affected the contact angle reading. To reduce this error, the samples were fixed to a holder which was moved up and down (1 cm) when measurements were taking place. This method allowed all samples to touch the water surface at consistent speed and angle (perpendicular to the water surface). Values of standard errors were reduced to some extent. Better results could be achieved if an automated sample holder and more powerful magnification devices were utilised.

Tensile test

Measurements of the mechanical properties of the membranes were performed with tensile tests. The material was clamped at both ends and stretched with constant speed. The force required to elongate the material with time was recorded until the material breaks. Parameters such as breaking force, breaking stress, maximum elongation and Young's modulus could be calculated using the obtained information. The tests were performed on a TA.XT2i/5 texture analyser (Texture Technologies) which has a maximum pulling force of 50N. The sample attachment adopted for testing the membranes is the tensile grip attachment procedure.

Flat sheet samples were cut into 3 x 1 cm and were soaked in Mili-Q water overnight to wet the samples before testing. The samples were placed in between grips, and preliminary tests were carried out to observe the behaviour of the sample. To obtain accurate results, the sample should break in the middle section of the grips. However, it was found that the samples broke at the clamped ends which indicated that the line grips pressed against the membrane and weaken the sample at this point.

It has been reported that grips lined with thin-rubber, crocus-cloth or pressure sensitive tape as well as file-faced or serrated grips have been successfully used for many materials (ASTM D882). In the study, both grips were lined with 4 pieces of 2 x 1cm sponge cloth (70% cellulose acetate & 30% cotton) before testing, 1 piece for each side of the grip. The same sponge cloth was used for all samples. Preliminary tests showed that this method is effective in breaking the samples at the middle section, an indicator that the samples were not weakened by the grips, as seen in Figure - A 3.

Figure - A 3: Tensile test sample testing with sponge cloth at the grips.

Hollow fibre

Prior to testing, the common protocol for hollow fibre membranes was to pot the sample fibre in plastic tubing at both ends. Without potting, the clamps could weaken the fibre resulting in the breakage of membrane at the grip. Hollow fibre samples were cut into 4 cm length before being placed in between tubing with 3.5 cm distance between ends. Epoxy resin was mixed and injected into the tubes using a syringe and the glue was ensured to fill the entire tube. The glue was allowed to set for 24 h. The samples were then immersed in Mili-Q water overnight. The length was measured again before testing as the effective length could change after potting.

Preliminary tests encountered several issues. Some of the samples broke at the potting point indicating that the epoxy resin could interact with and have an effect in degrading the strength of membrane material. For other samples, uneven resin distribution resulted in the membrane being "pulled out" of the potting tubes while being stretched by the machine before breakage occurs (Figure - A 4). Potting fouled HF/PP membranes that has been aged in NaOCl solution is difficult as the samples were very brittle and would break during membrane handling.

Figure - A 4: Potted hollow fibre sample being pulled out from the grips in tensile test.

To overcome the issues with potted membranes, tensile tests were run without sample potting. Both grips were lined with 4 pieces of 2 x 1cm sponge cloth as with the flat sheet samples. It was found that most of the samples broke in the middle section between the grips with no difference in results compared to successfully potted membranes. Fibre sample without potting was adopted as the sample preparation protocol for all hollow fibre membrane tested.

Tensile properties may vary with method of preparation, distance between the grips, specimen thickness, testing speed, types of grip used, manner of measuring extension, trigger load and distance (ASTM International). It is therefore important to be consistent in setting the testing parameter as protocol variations could generate different results.

New and aged membrane of the same material and configuration were tested with the same testing parameters to allow fair comparison between samples. The pre-test and post-test speed was fixed at 0.5 and 15 mm/s respectively. Other parameters were summarised in Table - A 1.

Table - A 1: Experimental parameters for tensile tests on different membrane types.

	Distance between grips (mm)	Fibre diameter / sheet thickness (mm)	Test speed (mm/s)	Distance (mm)
HF / PVDF	3.5	0.65	0.5	75
HF /PP	3.5	0.45	0.7	150
FS /PVDF	3.0	0.125	0.1	45

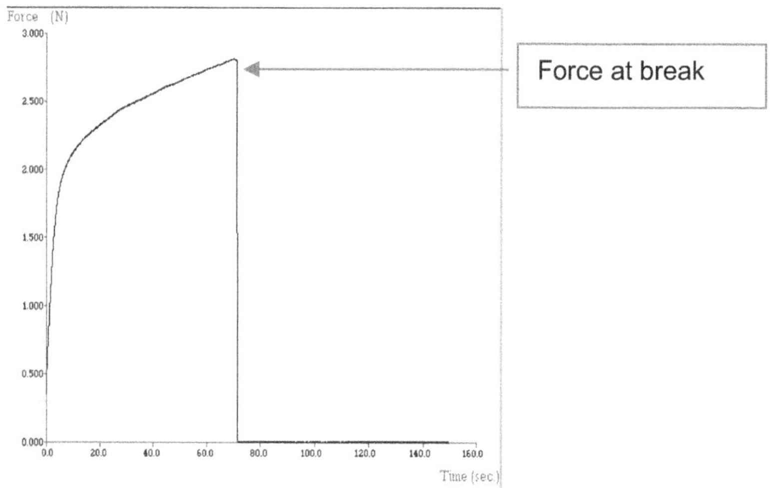

Figure - A 5: Information obtained from Stable Micro Systems® Texture Expert.

The software used to collect information from the Texture Analyser is Stable Micro Systems® Texture Expert. It recorded the force needed to elongate the material and the distance between grips per sampling time and plotted the graph of force (N) against time (s).The force at break could be observed as the vertical line when the force drops to zero (Figure - A 5). Further information on the physical properties and characteristics of the samples such as the stress-strain curve, breaking stress, maximum elongation and Young's Modulus could be drawn and calculated using the raw data collected.

Thermogravimetric analysis

Thermogravimetric analysis (TGA) is a thermal analysis technique which measures the weight change in a material as a function of temperature and time in a controlled atmosphere. The technique is useful is investigating the behaviour and thermal stability of materials, as well as performing compositional analysis of materials. Qualitative and quantitative analysis could be carried out with minimal solid sample size, with measurement range between 0.1-200mg.

TGA tests in this study were performed on the Thermal Advantage® device manufactured by TA Instruments. The samples were dried prior to testing to prevent the presence of moisture in the samples. Analysis was carried out in inert atmosphere (N_2 gas), with temperature increase in the range of 20-750°C. Test method chosen was the Ramp method, which increases the temperature 20°C/min.

Appendix 2

Accelerated Membrane Ageing Study
Report No. 3

Pierre LeClech

UNESCO Centre for Membrane Science and Technology

EXECUTIVE SUMMARY

Microfiltration (MF) and ultrafiltration (UF) have been increasingly used to remove pathogens, concentrate food products, and treat water and wastewaters. Whether using flat sheet or hollow fibres, many membrane filtration processes are subject to repetitive cycles of chemical cleanings to remove the materials retained on the membrane surface during operation. As cleaning agents generally remove efficiently deposited materials, those chemicals can also impair the membrane characteristics and selectivity, which can result in lower product quality and potential loss of membrane integrity. Although degradation of polymeric material has been widely studied in the literature, little is known about the nascent and cumulative changes in the chemical and structural composition of porous membranes due to chemical exposure. The methodology developed in this research is expected to provide a systematic assessment of membrane ageing, using previously recommended characterisation techniques. An assessment of the relative maximum cleaning intensity which can be imposed to porous membranes before significant change in membrane properties has been attempted in this study. The membrane ageing mechanisms were then studied by comparing and interpreting the different membrane characteristics.

Various membrane materials, as well as different configurations (hollow fibre and flat sheet) were applied to assess the cleaning tolerance of common porous membranes. The three hollow fibre membranes used in study were polyvinylidene fluoride (PVDF) and polypropylene (PP) and polyethersulfone (PES). The two flat sheet membranes were both PVDF membranes with hydrophilic (PVDF-HPI) and hydrophobic (PVDF-HPO) characteristics. Four chemical agents, commonly used in the water and wastewater industry, were selected for membrane ageing experiments: sodium hypochlorite (NaOCl), sodium dodecyl sulphate (SDS), hydrochloric acid (HCl) and sodium hydroxide (NaOH). The ageing process was carried out by soaking the wetted membranes in the cleaning solution for a maximum duration of 30 days. Following the ageing process, membrane samples were characterised using contact angle, thermogravimetric analysis (TGA), fourier transform infrared (FTIR) spectroscopy, tensile and bubble point tests.

The results obtained from the membrane characterisation allowed (1) an arbitrary ranking of membrane stability against different chemical cleaning agents (2) a better understanding of the mechanisms occurring during membrane ageing. Results indicated that PVDF-HF membrane properties were significantly modified after NaOCl and NaOH; whereas, SDS-aged membranes were not significantly affected. Both PVDF flat sheet membranes (PVDF-HPI and -HPO) were more sensitive to HCl exposure. Modifications to PP membrane properties were observed with NaOCl and HCl ageing equally. PES membrane was observed to be more sensitive to SDS exposure and, seemed to be slightly compromised after NaOH ageing.

All oxidising agents (NaOCl, HCl and NaOH) generally resulted in removal of membrane additives. For example, PVDF-HF membrane additives were removed by all three oxidising agents, although the extent of degradation of additives varied for different cleaning chemicals. NaOCl attack generally resulted in loss of membrane integrity through cross linking in PVDF, chain scission in PP, and the chlorination of PES. SDS cleaning led to subtle changes in membrane surface characteristics, which resulted in more hydrophilic and negatively charged membrane and the increase of membrane apparent pore size. The tested membranes exposure to HCl only resulted in modification of membrane surface properties (hydrophilisation), despite removal of additives. NaOCl and NaOH both resulted in degradation of all membranes core material. Similar to NaOCl, NaOH resulted in significant chemical degradation of the tested membranes. NaOH exposure generally resulted in membrane degradation through mostly dehydrofluorination in PVDF and chain scission through hydrolysis in PP and PES.

CONTENTS

FIGURES

TABLES

ABBREVIATIONS

AFM	- atomic force microscopy
ATR-FTIR	- attenuated total reflection - Fourier Transform infrared spectroscopy
CA	- cellulose acetate
CIL	- cleaning in line
CIP	- cleaning in place
CLSM	- confocal laser scanning microscopy
EDTA	- ethylene diamine tetra acetic acid
FS	- flat sheet
GPC	- gel permeation chromatography
HF	- hollow fiber
IC	- inorganic carbon
LMH	- Lm-2h-1
MBR	- membrane bioreactors
MF	- microfiltration
NaOCl	- sodium hypochlorite
PAN	- polyacrylontrile
PE	- polyethylene
PEG	- polyethylene glycol
PES	- polyethersulfone
PP	- polypropylene
PS	- polysulfone
PVDF	- polyvinylidene fluoride
PVP	- polyvinylpyrrolidone
Rm	- membrane hydraulic resistance
SDS	- sodium dodecyl sulphate
SEM	- scanning electron microscopy
TC	- total carbon
TGA	- thermogravimetric analysis
TMP	- trans membrane pressure
TOC	- total organic carbon
UF	- ultrafiltration
WT	- water treatment
WWT	- wastewater treatment
XPS	- X-Ray photoelectron spectroscopy

1. INTRODUCTION

Microfiltration (MF) and ultrafiltration (UF) have been increasingly used to remove pathogens, concentrate food products, and treat water and wastewaters. Whether using flat sheet or hollow fibres, many membrane filtration processes are subject to repetitive cycles of chemical cleanings to remove the materials retained on the membrane surface during operation. Even though cleaning agents can efficiently restore the flux, they can also impair selectivity of the membrane. Chemical cleaning has been reported to result in ageing of polymeric membranes [1-4] and a potential reduction of membrane lifetime and loss of integrity of separation process. Notwithstanding the importance of increasing membrane lifetime to improve the economic efficiency of the process, it is also critical to better detect nascent changes in the membrane characteristics before the quality of permeate gets affected.

Chemical cleaning agents are usually categorised according to their functional groups or types of foulant typically removed. Among the main organic foulants, proteins, polysaccharides, and humic substances are commonly found in both water and wastewater. Some of the main chemicals used in the water and wastewater treatment industry include alkali (caustics), oxidants and acids. Some of the main categories are listed in Table 1 together with typical foulants removed, their major cleaning mechanisms and typical chemicals used.

Table 1- Main categories of chemical cleaning agents and their properties [5-7].

Category	Typical targets	Mechanisms	Examples
Alkali	Proteins, carbohydrates, fats	Hydrolysis, solubilisation, saponification	NaOH
Oxidants/Disinfectants	Pathogenic microorganisms	Oxidation	NaOCl, H_2O_2, Peroxyacetic acid
Acids	Inorganic salts, oxide films	Solubilisation, hydrolysis	Citric, nitric, hydrochloric, phosphoric acid
Chelants	Mineral deposits	Chelation	Citric acid, ethylene diamine tetra acetic acid (EDTA)
Surfactants	Fats, proteins, oil	Dispersion	Surfactants (SDS), detergents
Enzymes	Proteins, lipids	Hydrolysis	Proteases, Lipases

Compatibility between membrane material and chemical agents, pH and operating conditions is important to preserve the lifespan of the membrane. pH and chemical resistances of different polymeric membranes are summarised in Table 2.

Table 2- Properties and recommended pH range of common polymeric materials used for MF and UF [8].

Material	pH range	Advantages	Disadvantages
PVDF	< 11	Acid resistant Chlorine resistant (high) Properties can be modified	Limited pH range
PES, PS	<13	Properties can be modified High pH range	
PE, PP	<14	Wide chemical resistance Chlorine resistant (moderate) Acid and caustic tolerant	Susceptible to oxidation Limited blend capacity

PS – polysulfone, PES–polyethersulfone, PVDF – polyvinylidene fluoride and, PP – polypropylene.

From the membrane degradation results compiled from previous works, it was observed that most research have focused on the effects of chemical agents on PS/PES and PA materials rather than on PVDF membranes. However, this material remains one of the most widely used types of polymer for membrane applied in the water industry today. Many studies have reported the effects of hypochlorites on membranes, and little information is available on the effects on other common chemicals (caustics, citric acid and surfactants) used in the industry. Further research is needed to better identify the membrane ageing effects of these chemicals, individually and in combination, on a wide range of membranes. Moreover, although polymeric degradation has been widely studied in the literature, little is known on the nascent and cumulative changes in the chemical and structural composition of porous membranes due to the action of chemical exposure.

This report focuses on the ageing effect of chemical agents (both oxidising and non-oxidising) commonly used in porous membrane plants. By using a wide range of analytical techniques - mechanical (tensile testing), morphological (bubble point pressure test and clean water test) and chemical (zeta potential analyser, thermogravimetric analysis and Fourier transform infrared spectroscopy (FTIR)), the complete assessment of the membrane state will be carried out before and after ageing. The pool of data obtained will allow (1) a better understanding of the mechanisms occurring during membrane ageing and (2) an arbitrary ranking of membrane stability against chemical agents.

2. MATERIALS AND METHODS

2.1 Membranes

Experiments were based on three commercially available micro-porous hollow fibre polyvinylidene fluoride (PVDF) and polypropylene (PP) and polyethersulfone (PES) membranes, and two PVDF flat sheet membranes of different characteristics (hydrophobic HPO and hydrophilic HPI). Both PVDF and PP hollow fibre membranes were supplied by Siemens. PES hollow fibre and flat sheet PVDF membranes were purchased from Koch membrane systems and Millipore, respectively. The properties of the membranes used in this study are presented in Table 3.

Table 3- The properties of the different membranes used in the accelerated ageing study.

ID	Core material	Configuration	Pore size (μm)	Outside diameter (mm)	Other
PVDF-HF	PVDF	Hollow fibre	0.04	1.30	-
PP	PP	Hollow fibre	0.20	0.45	-
PES	PES	Hollow fibre	0.05	2.60	-
PVDF-HPI	PVDF	Flat sheet	0.22	-	Hydrophilic
PVDF-HPO	PVDF	Flat sheet	0.22	-	Hydrophobic

"-": Not available

To prepare the membrane samples for accelerated ageing tests, 48 pieces of each hollow fibre and flat sheet membrane were cut into 30 and 5 cm x 5 cm, respectively. To wet these samples, all membranes were soaked in 80% ethanol for approximately 3 h. Then, all membranes were rinsed in a container filled with Milli-Q water for 3 days (stored at 4 °C), to remove the ethanol residue from the membranes. The rinsing water was replaced with fresh Mill-Q every 24 h.

2.2 Cleaning solutions

Four oxidising and non-oxidising chemical cleaning agents were prepared at different concentrations (Table 4).

Table 4- Cleaning agents concentration and their minimum and maximum cleaning intensities.

Cleaning agent	Concentration		Day 1 (ppm.h)	Day 30 (ppm.h)
	w/v %	ppm		
Sodium hypochlorite (NaOCl)	2	20,000	0.48×10^{6}	14.40×10^{6}
Sodium dodecyl sulphate (SDS)	2	20,000	0.48×10^{6}	14.40×10^{6}
Hydrochloric acid (HCl)	0.04	400	0.96×10^{4}	28.80×10^{4}
Sodium hydroxide (NaOH)	0.13	1,300	0.31×10^{5}	9.36×10^{5}

In the study, commercial NaOCl was sourced from Ajax Finechem with nominal concentration of 12.5% w/v. SDS, HCl and NaOH were purchased from Sigma Aldrich. Due to the aggressive nature of some these chemicals, different concentrations were used. For example, HCl is generally used in low concentrations since is highly corrosive and can lead to failure of the operating system. On the other hand, PVDF membrane material was found to be sensitive to high concentrations of NaOH, as the membrane colour changed from white to black immediately after exposure to 2% NaOH solution. Therefore, lower concentrations of HCl and NaOH were applied in this study.

2.3 Accelerated membrane ageing experimental conditions

In these experiments, pieces of each type of membranes were submerged in cleaning solutions. One batch of membrane and cleaning agents were used for better consistency. The solutions were refreshed every 3-4 days. Membrane samples were collected at 0, 1, 2, 3, 4, 7, 9, 10, 15, 20, 25 and

30 days from the start of the experiment. Then, membrane samples were dried in a desiccator for at least 24 h.

The hollow fibre membranes for clean water test and Bp test were potted in a loop shape using a PVC tube and Araldite to allow filtration. The potted samples were first wetted for 3 h in 80% ethanol solution and then rinsed with Milli-Q prior to ageing in the cleaning solutions.

2.4 Membrane characterisation

Following the ageing process, different membrane characterisation techniques were adopted to investigate the suitability of each method in assessing membrane degradation. In the previous study (Work Package 2), the suitability and reliability of the different membrane characterisation techniques in detecting membrane degradation were assessed and compared. The comparison was performed by developing a ranking system, with each criterion being given a weighting factor depending on its importance.

Table 5 shows the calculated reliability of the selected membrane characterisation techniques. These tests were carried out on five new samples of 0.04 µm HF/PVDF membranes. The average value of the samples was calculated for each test. The sample standard deviations, (SD), of each characterisation method were then computed to assess the variability in results from the mean value (see report #1 for more details).

Table 5- Summary of the reliability assessment of analytical techniques used on new PVDF membrane.

Characterisation techniques	Standard deviation (%)	
	Hollow fibre	Flat sheet
Clean water	6	19
Bubble point	20	-
Contact angle	20	3
Tensile	10	10
TGA (Temperature of the maximum weight loss, T_{max})	1	3

"-": Not available

2.4.1 Clean water test

Clean water test was performed after the wetting and rinsing procedure of the membranes to measure the intrinsic resistance of the membrane (R_m). For flat sheet membranes, filtration with Milli-Q water was carried out for 5 min to obtain pure water flux, J_0. The filtration was operated at constant trans-membrane pressure (TMP) of 20, 30 and 40 kPa and data was collected every second. Information recorded by the software include the pressure, weight of permeate collected and the calculation for flux at fixed sampling times.

The clean water test for the submerged hollow fibre membranes was performed using the flux stepping method. TMP was calculated as the difference in pressure between the filtrate and the permeate sides of the membrane. Constant flux filtration was carried out at 20 $Lm^{-2}h^{-1}$, and increased to 50 and 100 $Lm^{-2}h^{-1}$ after 30-60 s intervals. The pressure was recorded for each flux step. The R_m values were calculated from Equation 1.

$$R_m = \frac{TMP}{\eta J} \qquad \text{Equation 1}$$

Where η and J are viscosity (Pa.s) and flux ($Lm^{-1}h^{-1}$), respectively.

2.4.2 Bubble point test

For flat sheet membranes, the membrane was inserted in a high pressure dead-end cell. First 50 mL of Milli-Q was filtered through the membrane to ensure that membrane pores were filled with water.

Then, the permeate tube of the dead-end cell was submerged in a full beaker. During the test, the pressure was gradually increased until the first bubble was observed to appear out of the submerged permeate tube.

To determine the bubble point (Bp) pressure of the hollow fibre membranes, the potted modules were completely submerged in Milli-Q water, nitrogen gas was connected to the permeate side of the module. Then, the pressure was gradually increased until the first bubble was observed to leave the fibre surface.

2.4.3 Contact angle

The contact angle measurement of the new and aged membranes was performed with the KSV 200 CAM goniometer, with Milli-Q water used as the reference solvent. The technique used for the flat sheet membrane in the study was the sessile drop method, while meniscus measurement technique was adopted for the hollow fibre membranes.

The goniometer utilised a sample stage, a syringe to apply a droplet of liquid (for flat sheet membranes), a light source to illuminate the droplet, and a set of optics for magnifying the image for observation. The contact angle of the flat sheet membranes were determined by inserting 10 droplets on different locations of the membrane samples. The contact angle values were then determined by the software, and an average of the left and right angles of the droplet were calculated.

The recorded contact angle images for hollow fibre membranes needed to be manually adjusted. From the curve fitting window, the liquid baseline, meniscus/air boundary and solid/air boundary were fixed. The contact angle values were then calculated by the software using the average of the left and right meniscus values.

2.4.4 Tensile test

Measurements of the mechanical properties of the membranes were performed with tensile tests. The material was clamped at both ends and stretched with constant speed. The force required to elongate the material with time was recorded until the material breaks. The tests were performed on a TA.XT2i/5 texture analyser (Texture Technologies), which has a maximum pulling force of 250 N.

All samples were soaked in Milli-Q water overnight in order to wet the samples before testing. Hollow fibre samples were cut into 40 mm length. The test speed and the initial distance were 0.5 mm/s and 35 mm, respectively. Flat sheet samples were cut into 3 x 1 cm and. the test speed and the initial distance was adjusted to 0.1 mm/s and 30 mm, respectively. The software used to collect information from the Texture Analyser is Stable Micro Systems® Texture Expert. It recorded the force needed to elongate the material and the distance between grips per sampling time and plotted the graph of force (N) against time (s).

2.4.5 Thermogravimetric analysis (TGA)

TGA measures the weight change in a material as a function of temperature and time in controlled atmosphere. The technique is useful is investigating the behaviour and thermal stability of materials, as well as performing compositional analysis of materials. Qualitative and quantitative analysis could be carried out with minimal solid sample size, with measurement range between 0.1 and 200mg.

TGA tests in this study were performed on the Thermal Advantage® device manufactured by TA Instruments. The samples were dried prior to testing to prevent the presence of moisture in the samples. Analysis was carried out in inert atmosphere (N_2 gas), with temperature increase in the range of 20-1000°C. Test method chosen was the Ramp method, which increases the temperature 20°C/min.

2.4.6 Fourier transform infrared spectroscopy (FTIR)

FTIR allows the detection of changes in the functional groups of the membrane by comparing membranes before and after exposure to cleaning agents. FTIR detects bonds between atoms in a molecule and can thus provide information about functional groups and chemical structures of the membrane. Each peak illustrates a specific type of molecular vibration: symmetric, asymmetric, twisting, wagging, bending or rocking. The FTIR spectra is analysed with the high resolution Hummel polymer and additives spectrum library available with the FTIR program. Bruker IFS66/S FT/IR Spectrometer was applied to obtain the spectra for new and aged membranes.

3. RESULTS AND DISCUSSIONS

3.1 Effect of membrane rinsing

Prior to accelerated ageing tests, the membranes were wetted and rinsed by soaking in 80% ethanol for 3 h and with Milli-Q water for 3 days, respectively. Wetted membranes were characterised and compared with virgin material. To investigate the impact of membrane preparation, the TGA and FTIR characterisation results from the virgin and the wetted membrane samples were compared, in the initial step of experiment.

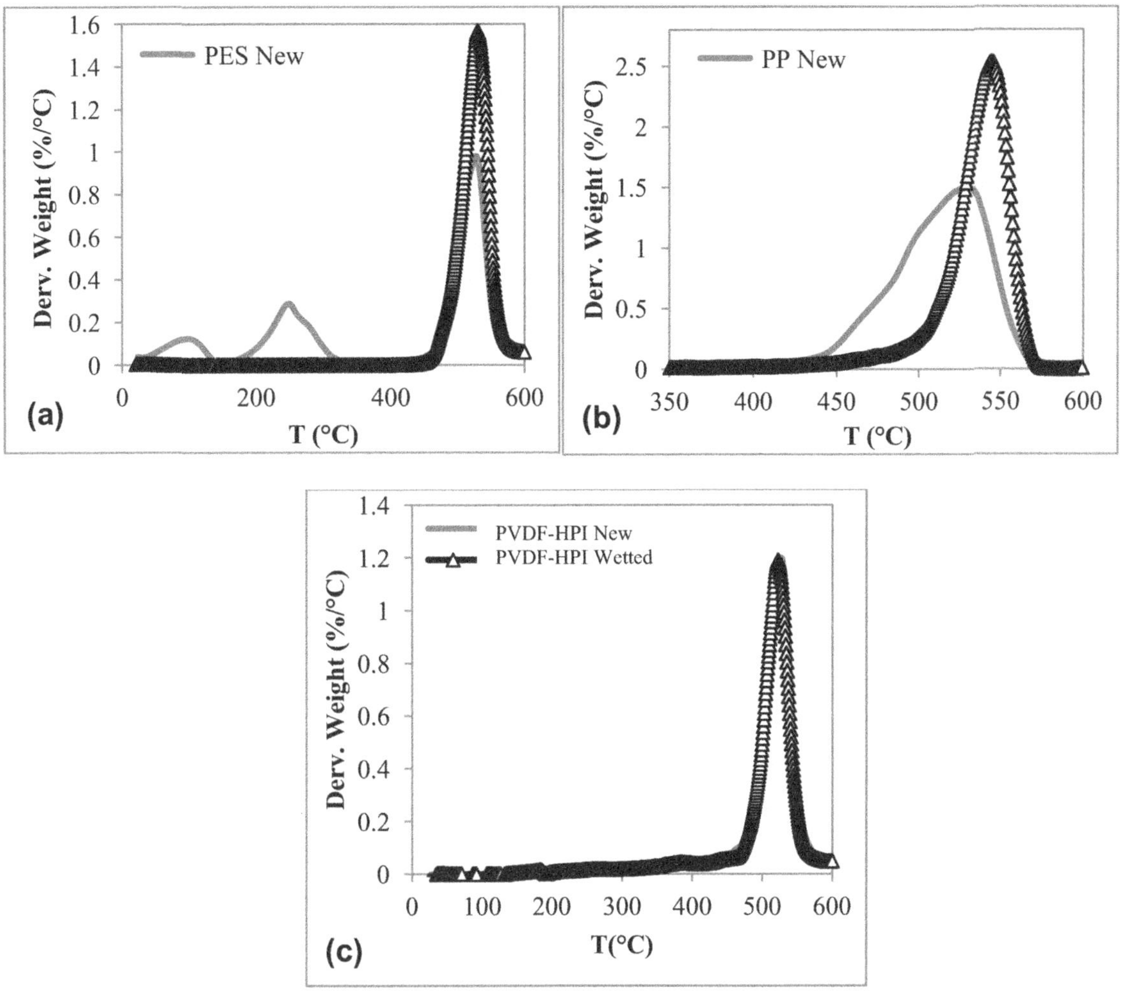

Figure 1- TGA curves for new and rinsed (a) PES, (b) PP and (c) PVDF-HPI membranes.

Figure 1 shows TGA results for PES, PP and PVDF-HPI flat sheet membranes. New PES membrane sample exhibited three stages of thermal degradation, which were reduced to one peak after wetting. The main peak corresponds to PES material itself, occurring in the range of 450-600°C with temperature of maximum mass loss rate (T_{max}) at 527°C. No significant characteristic changes in the main peak was observed after membrane wetting. However, TGA curve clearly shows two missing peaks at temperatures below 400°C. This most likely corresponds to preservative (for example, glycerine) that was removed during the wetting process. It is noteworthy that details of additives and preservatives used for commercial membranes are not generally revealed, sometimes proprietary. This restricts the comprehensive understanding of membrane degradation mechanisms and requires rigorous evaluations.

No distinctive preservative peaks were observed for the PP membrane. However, the peak observed for the new membrane was slightly, widened comparing to the wetted sample. This indicated that the degradation temperature of the preservatives used for PP membrane was close to PP itself and thus, these two peaks overlapped. Similar trends to PES were observed for PVDF-HF membrane.

New PVDF flat sheet membrane sample exhibited only one stage of degradation which corresponds to PVDF itself, occurring in the range of 480-580^{0}C with maximum degradation temperature at 526 ^{0}C. No significant characteristic changed at the second stage after membrane wetting. Therefore, it is most likely that the amount of preservatives in the new membrane sample was too small to be detected by TGA or that the preservative peak overlaps with the PVDF peak.

In the FTIR spectrum of the new and wetted PES membranes, indications of 1036 cm^{-1} (C-O stretch) peak suppression were observed after wetting, which does not belong to PES characteristic peaks [9]. This peak potentially originates from glycerine, i.e. preservative used during the PES membrane manufacturing process. Thus, missing and reduced peaks from FTIR spectra showed that the preservatives were removed due to wetting, confirming TGA results. FTIR study of the PVDF-HPI showed no significant change in the membrane (results not presented).

3.2 Accelerated ageing tests

In this study, five different membranes were aged with four cleaning agents, and sampled 7 times over 30 days. In order to limit the number of samples to be analysed, a ranking system was developed. During this process, all membrane samples collected on days 1 and 30 were analysed first, for all four cleaning agents. Then, the collected data were compared to those for the wetted membranes. The clean water test and the Bp analysis were carried out in reverse order in comparison with the other characterisation techniques applied in this study. The percentage of change of the average results for the aged (B) and the wetted (A) membranes was calculated according to Equation 2.

$$\text{Change \%} = \frac{|B-A|}{|A|} \times 100 \qquad \text{Equation 2}$$

If the percentage change in membrane characteristics was below 20 ± 2%, the maximum applied dosage (ppm.h) was expected not to alter that specific membrane property significantly. For change exceeding 20 %, the sample with the immediate lower cleaning intensity (i.e. day 20) was analysed. These trials were repeated until the percentage of change in that particular test results fell below 20 %.

Table 6 shows the TGA results obtained for wetted and aged samples (days 1 and 30). The thermal properties of these samples are expressed as percentage of mass loss and T_{max}. As mentioned previously, some of the membrane samples featured more than 1 stage of thermal degradation. The percentage of mass loss for each degradation stage of these samples is presented in Table 6. High reproducibility was observed for membranes, with average and maximum standard deviations less than 5%.

Table 6- TGA results for membranes aged in different cleaning solutions.

Membranes	Cleaning agent	Mass loss (%)		T_{max} (^{0}C)	
		Day 1	Day 30	Day 1	Day 30
PVDF-HF Mass loss = 22 and 47 % T_{max} = 544 °C	HCl	22 40	22 39	541	525
	NaOCl	13 63	9 54	544	526
	SDS	23 18	31 69	538	539
	NaOH	- -	19 56	-	520
PVDF-HPI Mass loss =72 % T_{max} = 544 °C	HCl	72	72	545	558
	NaOCl	82	80	572	553
	SDS	72	72	529	529
	NaOH	-	72	-	518
PVDF-HPO Mass loss =79 % T_{max} = 546 °C	HCl	75	76	533	554
	NaOCl	73	73	546	546
	SDS	76	84	532	527
	NaOH	-	63	-	528
PP Mass loss =100 % T_{max} = 543 °C	HCl	100	100	480	451
	NaOCl	100	99	538	535
	SDS	100	100	-	538
	NaOH	100	100	544	540
PES Mass loss = 77 and 7% T_{max} = 527 °C	HCl	66 10	64 10	510	484
	NaOCl	76 16	72 13	517	517
	SDS	- -	74 9	-	522
	NaOH	75 9	82 -	529	527

"-": Not available

Contact angle (CA) results for the accelerated membrane aged membranes are presented in Table 7. The standard deviation (SD) of these results was calculated from the triplicate analysis results (the SD values were calculated from the triplicate contact angle results).

Table 7- Summary of contact angle (CA) and standard deviation (SD) in ^{0} for aged membrane samples.

		PVDF-HF		PES		PP		PVDF-HPI		PVDF-HPO	
Membrane	Day	CA	SD	CA	SD	CA	SD	CA	SD	CA	SD
Wetted	0	28	3	18	1	28	4	85	4	118	2
HCl	1	19	2	12	2	8	2	58	4	133	2
	30	-	-	-	-	-	-	76	3	116	2
NaOCl	1	29	3	6	1	21	3	111	3	119	3
	30	28	3	-	-	22	2	109	6	125	2
SDS	1	18	1	9	3	21	2	66	2	132	5
	30	-	-	-	-	-	-	69	5	141	2
NaOH	1	19	1	13	1	14	3	85	4	116	4
	30	-	-	-	-	24	3	77	1	24	3

Information obtained from the experiment includes the fluctuations of flux and TMP at constant TMP for flat sheet at constant flux for the hollow fibre membranes respectively. The membrane resistance was calculated from the clean water test results, as presented in Table 8. The change % in resistance for most membranes already exceeded 20 % at day 15 and thus, those samples weren't aged further.

Hence, in Table 8, the membrane resistance at day 15 was reported instead of day 30. The SD values of the clean water test for the hollow fibre membranes was calculated from triplicate tests carried out. The SD value for the clean water test for hollow fibre membranes and the flat sheet membranes reported in Table 8 were 6 and 19% respectively. The high error in clean water test results was most likely due to the characteristics of these membranes or to the use of the different set-ups to carry out these experiments.

Table 8- Membrane resistance ($R_m \times 10^{-11}$) in m^{-1} for wetted and aged membranes in different cleaning solutions.

Membrane	Day	PVDF	PES	PP	HPI	HPO
Wetted	-	11.45	2.88	9.77	0.49	0.47
	1	10.07	3.42	12.42	-	-
HCl	15	6.51	0.63	2.42	0.54	0.55
	1	9.10	3.58	6.43	-	-
NaOCl	15	2.72	0.20	-	0.40	0.54
	1	8.22	3.71	7.02	-	-
SDS	15	3.21	1.36	0.65	0.46	0.63
	1	8.75	3.96	6.80	-	-
NaOH	15	5.54	0.80	2.30	0.65	0.45

"-": Not available

Following the clean water test, the samples were directly analysed to determine their Bp pressure (Table 9). Given its common use to assess membrane integrity in the industry, Bp tests were conducted within this project. However, the Bp test results for the virgin flat sheet membranes included a large error (± 50%) and therefore, were not reproducible enough to be reported. Also, since the Bp pressure of most membranes was observed to change above 20 % prior to 30 days of ageing, the obtained results for day 15 were presented in Table 9.

Table 9- Bubble point pressure (kPa) of the wetted and aged membranes (average SD of 20 kPa was obtained).

Membrane	Day	PVDF-HF	PES	PP
Wetted	-	240	100	200
	1	N/A	120	200
HCl	15	180	100	100
	1	200	40	200
NaOCl	15	180	20	N/A
	1	210	100	200
SDS	15	220	30	150
	1	180	110	200
NaOH	15	160	30	70

"-": Not available

The maximum elongation (E_{max}) data obtained from the tensile tests are reported in Table 10. The average SD of 9 and 13% were obtained for hollow fibre and flat sheet samples for the calculated E_{max} values, respectively. The tensile test was not possible for PES hollow fibre membranes since the force required to break the fibre exceeded the instrument limit (25 kg) due to the thick membrane support layer.

Table 10- Maximum elongation (E_{max} in %) of wetted and aged membranes in different cleaning solutions.

Membrane	Day	PVDF-HF	PP	PVDF-HPI	PVDF-HPO
Wetted	-	240	314	134	144
	1	237	358	125	141
HCl	30	251	289	130	131
	1	245	257	127	122
NaOCl	30	172	104	133	148
	1	262	278	135	140
SDS	30	234	289	133	125
	1	212	360	107	120
NaOH	30	201	290	124	133

"-": Not available

3.3 Ranking of membrane stability

3.3.1 Presentation of raw data

The results were then reported in terms of percentage of change between wetted and aged membrane (Equation 2). Based on the experimental plan used in this study, an estimation of the minimum cleaning intensity (CI_{min}) for which the membrane properties altered by 20 ± 2 % was possible. The cleaning intensity was calculated and reported in ppm.h. However, the CI_{min} value was estimated to sometimes be below the minimum or above maximum of the cleaning intensity tested ("< Min or "> Max" in tables). Table 11 shows the CI_{min} values determined where membrane modification was observed via TGA mass loss % alterations.

Table 11- Minimum cleaning intensity (CI_{min} in ppm.h) for which membrane modification was observed via TGA analysis (mass loss %).

	NaOCl	SDS	HCl	NaOH
Max tested	1,440,000	1,440,000	288,000	936,000
Min tested	480,000	480,000	9,600	31,200
PVDF-HF	< Min	<960,000	> Max	> Max
PVDF-HPI	> Max	<960,000	> Max	> Max
PVDF-HPO	> Max	> Max	> Max	> Max
PP	> Max	> Max	> Max	> Max
PES	> Max	> Max	> Max	> Max

From the results obtained through TGA analysis, only the mass loss % was used for ranking of membrane stability. Alterations in the mass loss% generally indicated that the membrane additives have been removed. However, the T_{max} value can also provide information regarding the degradation of the core material. The percentage of change was not reported due to fluctuation of the T_{max} values depending on the sample, and will be further discussed in Section (3.4).

As can be seen in Table 11, 480,000 ppm.h NaOCl exposure modified the PVDF-HF membrane mass loss in both degradation stages (i.e. the additives and the core material) significantly. Whereas, there were no significant changes in TGA mass loss observed within the 1,440,000 ppm.h for the other tested membranes. TGA results obtained for both PVDF-HF and -HPI membranes indicated that their thermal properties altered significantly, below 9,600,000 ppm.h SDS exposure. The maximum cleaning intensity of HCl did not seem to result in any modification in all tested membranes mass loss % during TGA analysis. Similar to HCl, NaOH maximum cleaning intensity did not feature any significant change on membrane samples mass loss %. This was due to the relatively lower cleaning intensity imposed during the test conducted with HCl and NaOH. Therefore, in this study, only NaOCl and SDS feathered an impact in terms of mass loss% of aged membrane samples during TGA analysis. The PVDF-HF appeared to be the most easily affected membrane. Since the mass loss %

reflected on the potential removal of the additives used to increase the hydrophilicity of the core material, it is expected that the degradation of surface modifiers were responsible for these results.

Table 12- Minimum cleaning intensity (CI_{min} in ppm.h) for which membrane modification was observed via contact angle analysis.

	NaOCl	SDS	HCl	NaOH
Max tested	14,400,000	14,400,000	288,000	936,000
Min tested	480,000	480,000	9,600	31,200
PVDF-HF	< Min	< 3,360,000	< Min	< Min
PVDF-HPI	< Min	< Min	< Min	< Min
PVDF-HPO	> Max	< 14,400,000	< Min	< 218,400
PP	< 3,360,000	< Min	< Min	< Min
PES	< Min	< Min	< Min	< Min

Although the TGA results did not show a significant impact during accelerated ageing of most membrane samples, contact angle results (Table 12) appeared to be more sensitive. For samples aged in NaOCl solution, only the contact angle of PVDF-HPO membrane was not altered with maximum cleaning intensity. HCl ageing on the other hand, resulted in a significant change at the CI_{min} for all the tested membranes.

Table 13- Minimum cleaning intensity (CI_{min} in ppm.h) for which membrane modification was observed via clean water test analysis.

	NaOCl	SDS	HCl	NaOH
Max tested	14,400,000	14,400,000	288,000	936,000
Min tested	480,000	480,000	9,600	31,200
PVDF-HF	< Min	<3,360,000	<19,200	< Min
PVDF-HPI	> Max	<2,400,000	< Min	<156,000
PVDF-HPO	< 7,200,000	<7,200,000	< Min	<218,400
PP	< Min	<3,360,000	< Min	<156,000
PES	< 960,000	< Min	< Min	<156,000

The CI_{min} for which the membrane resistance altered by 20±2 % are shown in Table 14. These results indicate that all membranes, except for HPI, experienced a significant change in their resistance below the maximum cleaning intensity level. These alterations were more significant for HCl and NaOCl amongst all the cleaning agents. PVDF-HF was observed to be the most HCl resistant amongst all the tested membranes, as alteration in its resistance occurred in 19,200 ppm.h of HCl ageing. NaOCl attack was relatively less pronounced for PVDF flat sheet membranes. On the other hand, the impact of SDS on the R_m of all samples except PES was observed be relatively less severe.

Table 14- Minimum cleaning intensity (CI_{min} in ppm.h) for which membrane modification was observed via bubble point (Bp) test analysis.

	NaOCl	SDS	HCl	NaOH
Max tested	14,400,000	14,400,000	288,000	936,000
Min tested	480,000	480,000	9,600	31,200
PVDF-HF	< 7,200,000	< 960,000	<144,000	< Min
PP	< 960,000	< 7,200,000	< Min	<218,400
PES	< Min	< 2,400,000	<192,000	<64,200

Table 14 shows the CI_{min} for which the Bp pressure of the membranes altered. Again, all membranes showed a significant change within the maximum cleaning intensity limit imposed by all the cleaning agents used in this study.

The FTIR analysis of the aged membranes showed that most samples chemical characteristics altered after the membranes were exposed to the cleaning agents (Table 15). These changes were generally observed for the carbonyl peaks (1650 and 1734 cm^{-1}) which represent membrane additives in wetted membranes. Also, in some cases new peaks emerged in the aged membrane spectra. It is noteworthy that the emergence or suppression of peaks was reported as the minimum cleaning dose here, even though the intensity of these changes seemed small. This was due to the lack quantitative analysis by FTIR. However, the FTIR spectra are presented and discussed more in detail in the following sections.

These alterations were more significant for NaOCl and NaOH, amongst all the cleaning agents. NaOCl attack was relatively less pronounced for PES and HPO membranes. The effect of NaOH on PVDF membranes both hollow fibre and flat sheet was more significant than PES. On the other hand, SDS and HCl were observed to effect PVDF flat sheet membranes only.

Table 15- Minimum cleaning intensity (CI_{min} in ppm.h) for which membrane modification was observed via FTIR analysis.

	NaOCl	SDS	HCl	NaOH
Max tested	14,400,000	14,400,000	288,000	936,000
Min tested	480,000	480,000	9,600	31,200
PVDF-HF	< Min	> Max	> Max	< Min
PVDF-HPI	< Min	< Min	< Min	< Min
PVDF-HPO	<3,360,000	<9,600,000	< Min	< Min
PP	< Min	< Min	> Max	> Max
PES	<3,360,000	< Min	> Max	< 936,000

The minimum cleaning intensities in which modification in maximum elongation of the aged membranes in different cleaning solutions were observed, are presented in Table 16. As mentioned previously, the force required to break the PES fibres exceeded the texture analysers maximum capacity and thus, PES aged samples could not be analysed here. As can be seen in Table 16, SDS and HCl did not significantly impact on these membranes mechanical properties. On the other hand, NaOCl exposure compromised PP and PVDF membranes maximum elongation properties. This impact was observed to be more severe for PP, as the membrane modification occurred during a lower cleaning intensity than PVDF. Furthermore, the maximum cleaning intensity of NaOH resulted in a significant modification of PVDF tensile properties.

Table 16- Minimum cleaning intensity (CI_{min} in ppm.h) of which membrane modification was observed in maximum elongation (E_{max}) via tensile analysis.

	NaOCl	SDS	HCl	NaOH
Max tested	14,400,000	14,400,000	288,000	936,000
Min tested	480,000	480,000	9,600	31,200
PVDF-HF	<9,600,000	> Max	> Max	<936,000
PVDF-HPI	>Max	> Max	> Max	> Max
PVDF-HPO	> Max	> Max	> Max	> Max
PP	< Min	> Max	> Max	> Max

3.3.2 Interim conclusions

The objective of this section is to assess the relative impact of the different membrane cleaning agents on membrane modification. The comparison was performed by developing a ranking system, in which, observed changes via each characterisation technique at the CI_{min} applied was evaluated. An analysis was carried out using points system to compare the membrane properties modification through each membrane characterisation technique. Where "✓" indicates that no modification was observed, "×" was used for significant change and "-" for indicated that the results for te given membrane /testing was not available.

From Table 17, it was observed that PVDF-HF membrane properties were modified at the CI_{min} of NaOCl and NaOH. On the hand, SDS ageing did not result in any significant modification of PVDF-HF membrane. Both PVDF flat sheet membranes were more sensitive to HCl exposure. Modifications to PP membrane properties were observed with NaOCl and HCl ageing equally. PES membrane was observed to be more sensitive to SDS exposure and, seemed to be less compromised after NaOH ageing.

Table 17- Summary of impact of different cleaning agents on various membrane properties. Results are given in this order: TGA, CA, R_m, Bp, FTIR and E_{max} for CI_{min}.

	NaOCl	SDS	HCl	NaOH
CI_{min} (ppm.h)	480,000	480,000	9,600	31,200
PVDF-HF	×××✓×✓	✓✓✓✓✓✓	✓×✓✓✓✓	✓××××✓
PVDF-HPI	✓×✓-×✓	✓×✓-×✓	✓××-×✓	✓×✓-×✓
PVDF-HPO	✓✓✓-✓✓	✓✓✓-✓✓	✓××-×✓	✓✓✓-×✓
PP	✓✓×✓××	✓×✓✓×✓	✓×××✓✓	✓×✓✓✓✓
PES	✓×✓×✓-	✓××✓×-	✓××✓✓-	✓×✓✓✓-

(×:significant change ✓:no change -:result not available)

As expected, SDS ageing resulted in less stringent modifications in most membrane, where as NaOCl and HCl ageing significantly compromised their properties. However, whether or not if these changes were related to the degradation of the core membrane material cannot be assessed through the ranking system applied here. The following sections will discuss the membrane degradation in more detail.

3.4 Membrane degradation mechanisms during accelerated chemical cleaning

3.4.1 Assessment of membrane degradation by NaOCl

NaOCl is one the most common oxidants used for membrane chemical cleaning, although it is expected to participate to the degradation and the loss of membrane integrity [3, 10-12]. As mentioned in the previous section, alteration of all the tested membranes was observed as a result of NaOCl cleaning. Furthermore, the minimum intensity for which the membranes were modified as the result of NaOCl exposure varied significantly for different membrane materials and characterisation methods. In this section, the possible degradation mechanisms occurring during NaOCl attack were assessed for all the tested membranes.

3.4.1.1 PVDF-HF

Figure 2 shows the properties of wetted and aged PVDF-HF membranes. The TGA results indicated that the peak observed in the wetted sample between 360 and 480 °C (24% mass loss), corresponding to the thermal degradation of polyvinylpyrrolidone (PVP, i.e. a common membrane additive) which was gradually removed during the ageing process. Additives in casting solution are widely used for pore-forming and/or surface modification, in order to create a more hydrophilic membrane [13-14]. The minimum cleaning intensity resulted in partial removal of surface modifiers added to the PVDF-HF blend. The degradation of PVP continued further for higher cleaning intensities, until it was removed completely from 9.60×10^6 ppm.h.

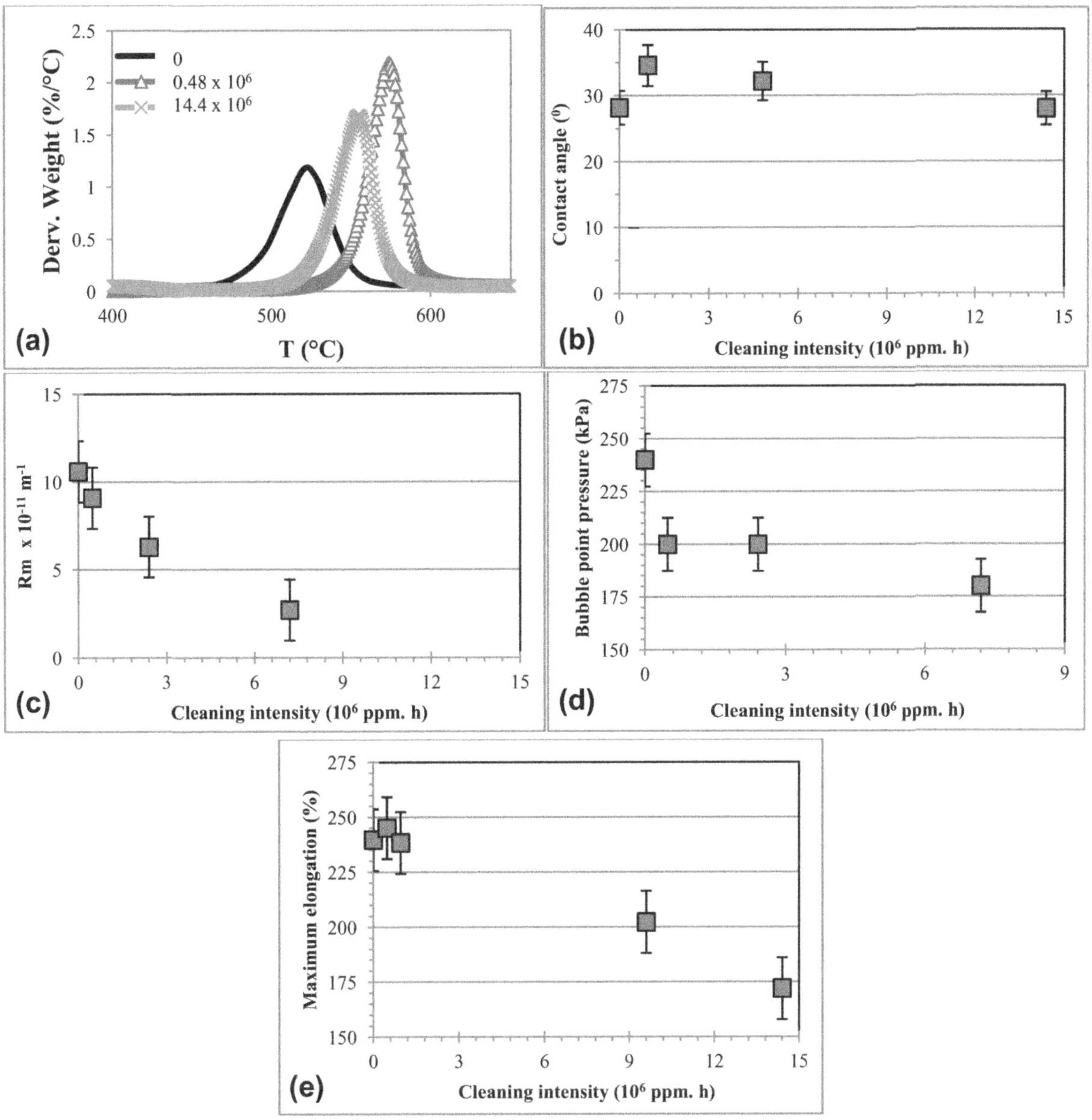

Figure 2- (a) TGA, (b) contact angle, (c) membrane resistance, (d) bubble point pressure and (e) maximum elongation of PVDF-HF membrane aged with 2% NaOCl.

Moreover, the T_{max} for PVDF-HF aged membrane was observed to shift to a higher value. Then, it slightly decreased, as the membrane was further exposed to the cleaning agent. The trends observed for the PVDF peak can be due to two common degradation mechanisms, which are cross linking and chain scission [15-16]. In this case, it is likely the T_{max} first increased as the result of cross linking, before gradually decreasing as the polymer became less stable due to occurrence of chain scission. Therefore, it can be proposed that, membrane degradation first started with cross linking between the polymers chains, before the higher cleaning intensity initiated further membrane ageing through chain scission.

According to

Figure 2-(b), the membrane contact angle slightly increased at the lower cleaning intensities. However, the membrane became more hydrophilic when it was aged above 4.8×10^6 ppm.h. The decrease in membrane Bp pressure and resistance indicated that pore size may have enhanced. However, surface modification of the PVDF membranes has been already reported after accelerated ageing with 1% NaOCl [3]. In the previous study, the membrane hydrophobicity increased significantly in the initial stage of ageing (up to approximately 6.72×10^6 ppm.h), before decreasing to a level lower than of the virgin membrane. Although the trends in the two studies are relatively similar, the high initial increase in the contact angle declined at a lower cleaning intensity ($4.8x10^6$ ppm.h) in the present research. This comparison suggested that, although ppm.h is used to express the cleaning intensity, the cleaning agent concentration was probably more influential than ageing time in their series of experiments. Besides the removal of additives, pore size enhancement was also reported in the later study [3], which confirmed the decreased Bp pressure observed here.

The FTIR analysis of the aged PVDF-HF (see Appendix A) was performed and the alterations in membrane function groups were analysed. Peak suppression was observed between 1660 to 1740 cm^{-1}, which represented the destruction of carbonyl group present in both the PVDF and PVP compounds. No further alteration in the membrane chemical structure was observed from the FTIR spectra.

Although the FTIR spectra support the TGA results regarding the elimination of additives, the occurrence of cross linking could not be confirmed. Generally, cross linking between the polymer chain results in additional carbonyl bonds, peak growing or broadening of the carbonyl region [17], could not be clearly observed here. This could be due to overlapping peaks in the FTIR spectrum. However, PVDF membranes are generally considered as one the most oxidant and chlorine resistant membranes compared to other polymeric membranes [18]. The results for the current study were confirmed by the trends reported by Puspitasari *et al.* (2010) that, NaOCl cleaning did not seem to significantly affect the membrane functional groups, apart from the disappearance carbonyl peak at 1729 cm^{-1}.

3.4.1.2 PVDF-HPI

The PVDF-HPI membrane properties are shown in Figure 3. As expected, the trends observed for PVDF-HPI (i.e. flat sheet) membrane were similar to those for PVDF-HF membrane. The TGA results showed the significant shift of the maximum weight loss rate to a higher temperature which was followed by a slight decrease as the cleaning intensity increased.

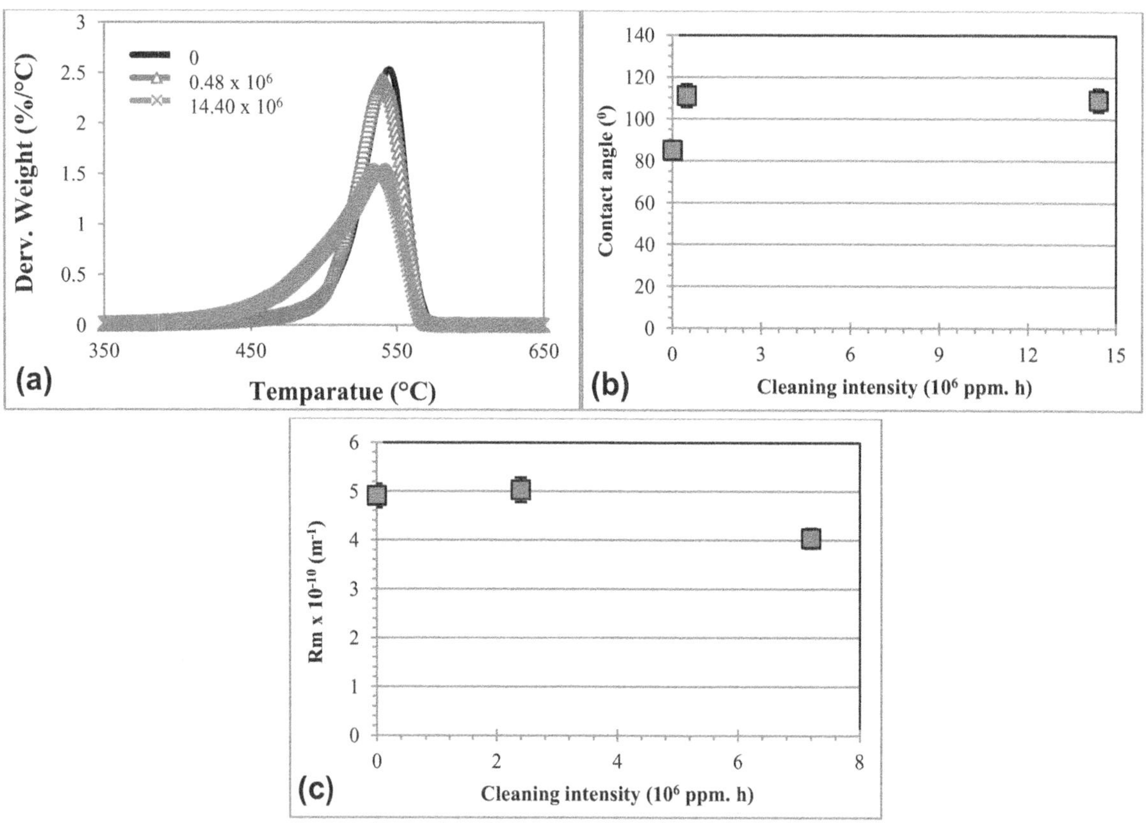

Figure 3- (a) TGA, (b) contact angle, (c) resistance of HPI membrane aged with % 2 NaOCl.

After initial ageing, the contact angle of HPI increased from 85 to 111 ^{0}; however, it did not decrease any further for higher cleaning intensities, as it was previously observed during the ageing of PVDF-HF membrane. The PVDF-HPI membrane was a PVDF membrane, which was specifically modified by the manufacturer to exhibit hydrophilic properties, since the PVDF is known to be rather hydrophobic. NaOCl cleaning has been reported to degrade the added surface modifiers [3]. The FTIR spectra of the aged PVDF-HPI membrane in NaOCl (Appendix A) indicated that the additives were completely removed from the membrane chemical structure.

3.4.1.3 PVDF-HPO

The PVDF-HPO membrane properties are shown in

Figure 4. The trends observed for PVDF-HPO was very similar to the ones for PVDF-HPI. However, the NaOCl attack to the PVDF-HPO membrane appears to be less aggressive, as the thermal degradation and the tensile properties of the PVDF-HPO membrane did not seem to be compromised, as the PVDF-HPO membrane was expected to contain less or no hydrophilic additives. The mechanism of degradation on PVDF-HPO was most likely to involve only cross linking. This hypothesis was confirmed by the developing two new carbonyl peaks observed in the aged PVDF-HPO membrane FTIR spectra (see Appendix A).

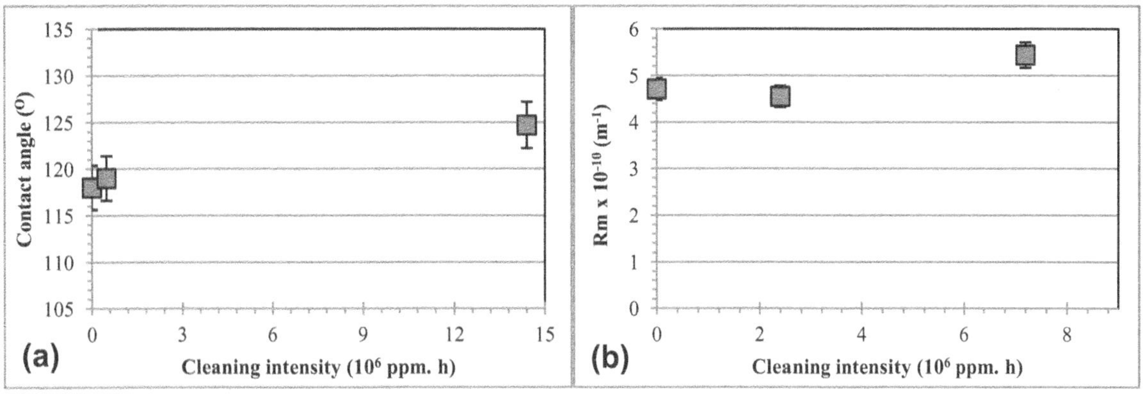

Figure 4- (a) Contact angle and (c) resistance of HPO membrane aged with % 2 NaOCl.

3.4.1.4 PP

PP membranes were reported to be less resistant to oxidants such as NaOCl when compared to other common polymers [19]. A significant compromise in PP membrane integrity was indeed observed during the accelerated ageing tests, as the membrane became completely brittle after 7.2 x 10^6 ppm.h of NaOCl exposure.

Figure 5 shows the properties of the wetted and the aged PP membranes in NaOCl solution.

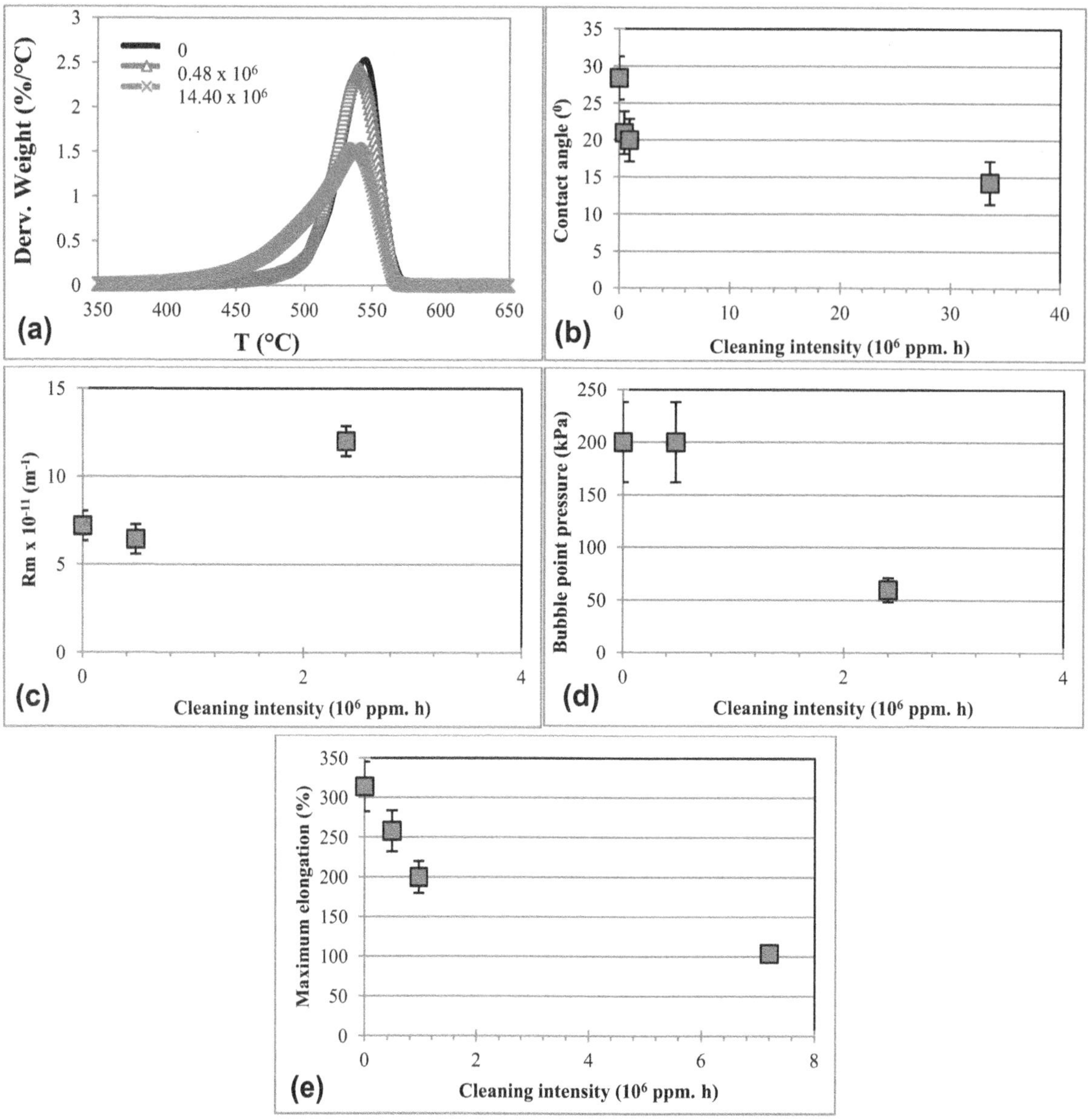

Figure 5- (a) TGA, (b) contact angle, (c) resistance, (d) bubble point pressure and (e) maximum elongation of hollow fibre PP membrane aged with 2% NaOCl.

As can be seen in

Figure 5-(a), the peak in the TGA graph for the PP membrane aged at the maximum cleaning intensity applied in this study (14.40 x 10^6 ppm.h) appeared wider than the other ones. This indicates that the thermal degradation temperature of PP membrane decreased, due to lower stability of the polymer as a result of ageing. Generally, decrease in thermal degradation temperature of polymers can be linked to occurrence of random chain scission.

The proposed ageing mechanism based on the TGA results was further confirmed by the observed loss in the PP membrane mechanical properties, as shown in

Figure 5-(e). The sharp decrease in maximum elongation of the PP membrane fibres (from 314 to 104 %) indicated the severity of the embrittlement occurring during NaOCl ageing.

In terms of membrane hydraulic performances, the R_m values were observed to slightly decrease in the initial stage of ageing. Then, the PP membrane resistance significantly increased with the accelerating cleaning intensity before it became completely brittle. Therefore, the clean water and the Bp tests were not conducted, since the membrane was too fragile for 7.20×10^6 ppm.h. Furthermore, significant decrease was observed in the membrane Bp pressure, indicating that the PP membrane was severely compromised.

On the other hand, the contact angle result (

Figure 5-(b)) showed that the PP membrane became more hydrophilic after the membrane ageing. This was in contradiction with the trend observed for membrane resistance, and can be due to the effect of pore size and surface roughness on the measured contact angle [20].

The FTIR spectra of the aged PP membrane in NaOCl (Appendix A) indicated significant alterations occurring in the membrane chemical structure. In the FTIR spectrum of the aged PP membrane, O-H groups emerged whereas, the C=O band disappeared. Moreover, the carbonyl peak (from 1653 to 1607 cm^{-1}) shifted to frequency of C-C band. These changes confirmed the reported chain scission degradation process of PP polymer. Therefore, FTIR analysis of the aged sample indicated that chain scission occurred due to PP membrane exposure to NaOCl, which confirmed the decrease in the degradation temperature observed during TGA analysis.

3.4.1.5 PES

PES has been considered resistant to oxidation and chlorine tolerant. However, degradation of PES was observed when it was exposed to high concentrations of NaOCl in a recent accelerated ageing study [20]. The contact angle of the PES membrane decreased from 18 to 6 ^{0}, when the membrane was aged for 0.48×10^6 ppm.h.

Figure 6 shows (a) resistance and (c) Bp pressure of PES hollow fibre membrane before and after ageing in 2 % NaOCl.

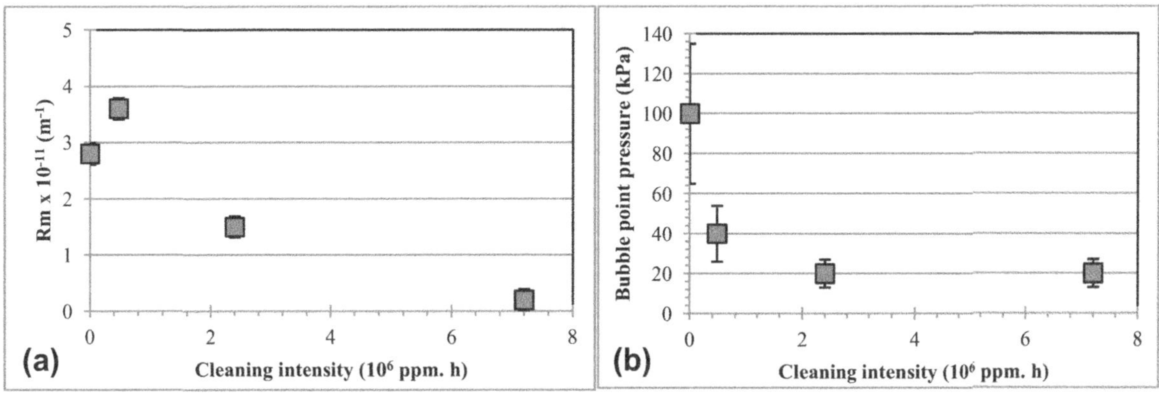

Figure 6- (a) Resistance and (b) bubble point pressure of hollow fibre PES membrane aged with % 2 NaOCl.

The membrane resistance trend showed a slight increase, followed by a sharp decrease as the cleaning intensity increased. The significant decrease in the membrane Bp pressure indicated that the membrane pore size enhancement has occurred, even in the minimum cleaning intensity (0.48×10^6 ppm.h).

The obtained results were confirmed by similar trends reported in the literature [20]. Aged PES flat sheet membrane with NaOCl was reported to feature a larger pore size, higher hydrophilicity and lower maximum elongation. The proposed degradation mechanism for PES in the latter study was reported to occur through chlorination of C-S bonds in PES. The FTIR analysis of the aged PES membrane (see Appendix A) indicated that the intensity of the peak at 1585 cm^{-1} corresponding to C-S band decreased which confirmed the chemical attack of NaOCl to the PES polymer. Therefore, in the case of PES membrane and NaOCl cleaning, the direct degradation of the core material was observed rather than subtle changes related to removal of additives.

3.4.2 Assessment of membrane degradation by SDS

SDS is an anionic surfactant with negatively charged hydrophilic groups, which can be used to remove nonionic organic compounds and divalent metal cations through emulsification, dispersion or surface conditioning. The observed alterations in the tested membranes properties and the possible ageing mechanisms as the result of SDS cleaning are discussed in this section.

3.4.2.1 PVDF-HF

The most significant changes in PVDF-HF membrane properties observed after the accelerated membrane ageing are presented in Figure 7. The results indicated that SDS impact on the PVDF-HF membrane was only significant on the membrane hydrophobicity. As can be seen in Figure 7(a), the membrane contact angle decreased significantly from 28 to 8 ^{0} due to SDS cleaning. With increasing cleaning intensity, the membrane continued to feature more hydrophilic properties. This observed trend was confirmed by the clean water test results, which decreased significantly at the minimum cleaning intensity for SDS (0.48 x 10^6 ppm.h). SDS features hydrophilic groups, which have the potential to attach to the membrane surface and thus, increase the membrane hydrophilicity and permeability [19].

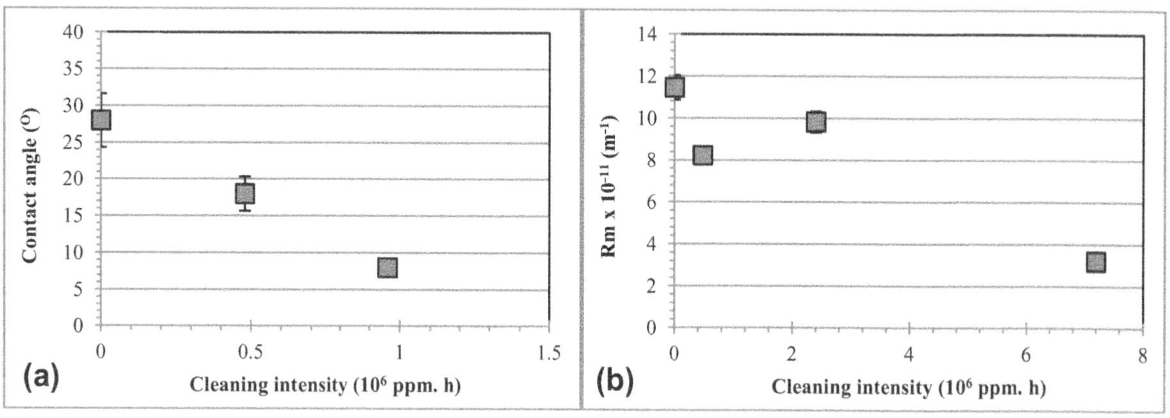

Figure 7- (a) Contact angle and (c) resistance of PVDF membrane aged with 2% SDS.

The FTIR analysis of the aged PVDF-HF membrane indicated that SDS cleaning did not affect the membrane additives (see Appendix B). Given that SDS did not possess any oxidative characteristics, alterations in the chemical structure of PVDF polymer was indeed not expected.

3.4.2.2 PVDF-HPI

As can be seen in Figure 8, SDS ageing seemed to affect the membrane additives from the PVDF-HPI membrane. The observed thermal degradation curve for the wetted PVDF-HPO started at approximately 460 ^{0}C, whereas for the 0.48 x 10^6 ppm.h aged sample it was at 483 ^{0}C. Moreover, the T_{max} of the PVDF-HPI membrane aged at the maximum cleaning intensity increased by 7 ^{0}C. However, this increase did not significantly exceed the SD of these measurements (1%). On the other hand, the PVDF-HPI membrane, after SDS ageing, became more hydrophilic, which was due to the hydrophilic nature of this cleaning agent.

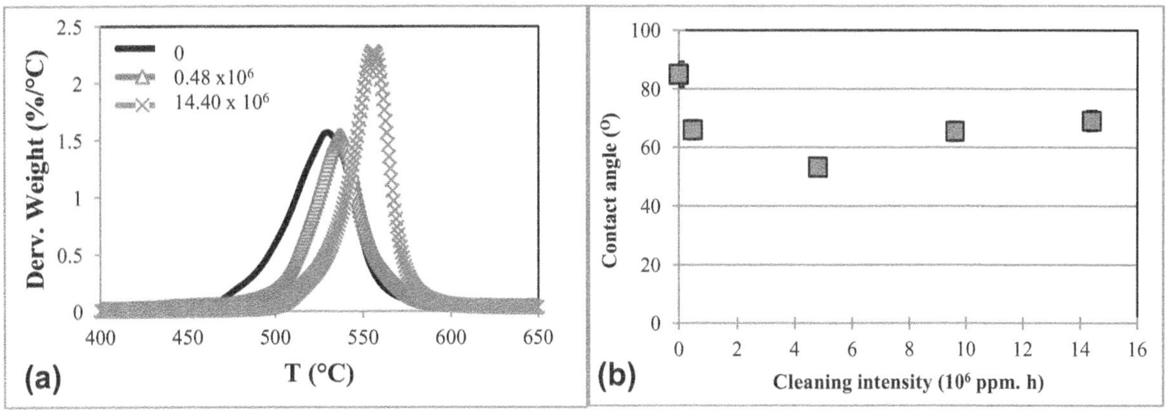

Figure 8- (a) TGA and (b) contact angle of PVDF-HPI membrane aged with 2 % SDS.

The FTIR spectrum of the SDS-aged PVDF-HPI membrane (see Appendix B) confirmed the removal of peaks at 1659, 1585 and 1489 cm^{-1}, corresponding to membrane additive (e.g. PVP). However, the other carbonyl peak at 1730 cm^{-1}, which also corresponds to another membrane additive, was not affected by SDS cleaning.

3.4.2.3 PVDF-HPO

In contrast to the all other PVDF membranes tested in this study, the PVDF-HPO membrane contact angle increased gradually after SDS ageing (Figure 9). The increased resistance of the membrane was consistent with the trend observed for contact angle of the PVDF-HPO membrane. The FTIR spectrum of the PVDF-HPO membrane for aged sample for 14.4 $x10^6$ ppm.h (see Appendix B) did not display alternations in the membrane chemical structure.

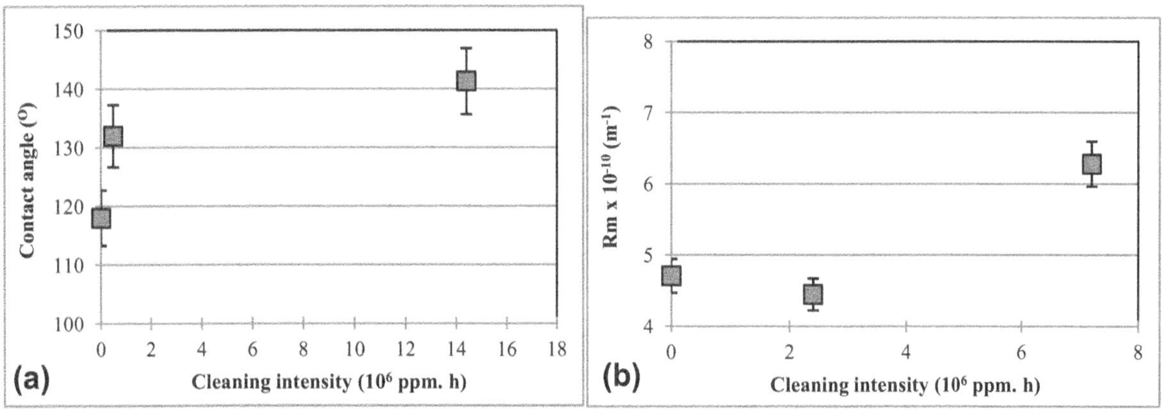

Figure 9- (a) Contact angle, (b) resistance PVDF-HPO membrane aged with 2 % SDS.

3.4.2.4 PP

The observed changes in PP membrane characteristics are shown in Figure 10 at the minimum SDS cleaning intensity. The PP membrane contact angle and R_m decreased. This decrease in membrane hydrophobicity led to a significant increase in membrane permeability, which was observed from the declining resistance trend. On the other hand, the observed decrease in PP membrane Bp pressure suggested that the decreased R_m (to $0.7x10^{11}$) was possibly not only due to the more hydrophilic membrane surface after SDS cleaning.

The FTIR spectra of the aged PP membrane (see Appendix B) at the maximum intensity of SDS cleaning did not show a significant difference compared to the wetted membrane. Therefore, the membrane chemical properties did not seem to alter due to SDS exposure.

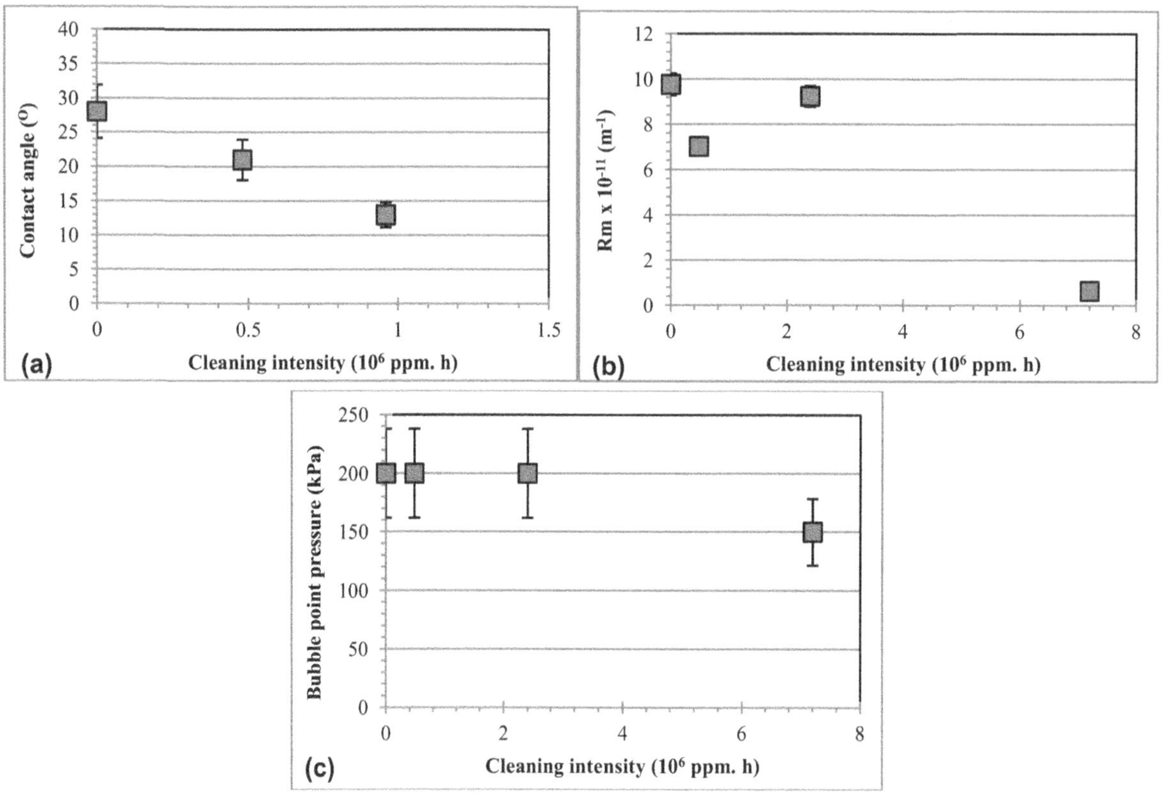

Figure 10- (a) Contact angle, (b) resistance and (c) bubble point of PP membrane aged with 2 % SDS.

SDS adsorption through hydrophobic interactions between the membrane surface and of the surfactant tails [21] was reported to increase the negative charge of DK nanofiltration (NF) membrane [22]. The decreased Bp pressure observed here can be due to opening of the membrane pores as the result of increased negative charge inside the pores of the SDS-aged membrane. As mentioned previously, SDS is a negatively charged hydrophilic surfactant which is generally difficult to remove from the surface; its functional groups can form hydrogen bonding with the membrane [19]. This residue of SDS on the membrane can increase the membrane surface charge and thus, lead to membrane swelling.

3.4.2.5 PES

Trends observed for the aged PES membranes were relatively similar to those for PVDF-HF and PP membranes. Again, decreases in PES membrane contact angle, resistance and Bp pressure were observed in Figure 11. However, at the minimum SDS cleaning intensity, the PES membrane contact angle and permeability both decreased significantly. At higher SDS cleaning intensity, PES contact angle increased; whereas, the membrane resistance was observed to decrease significantly.

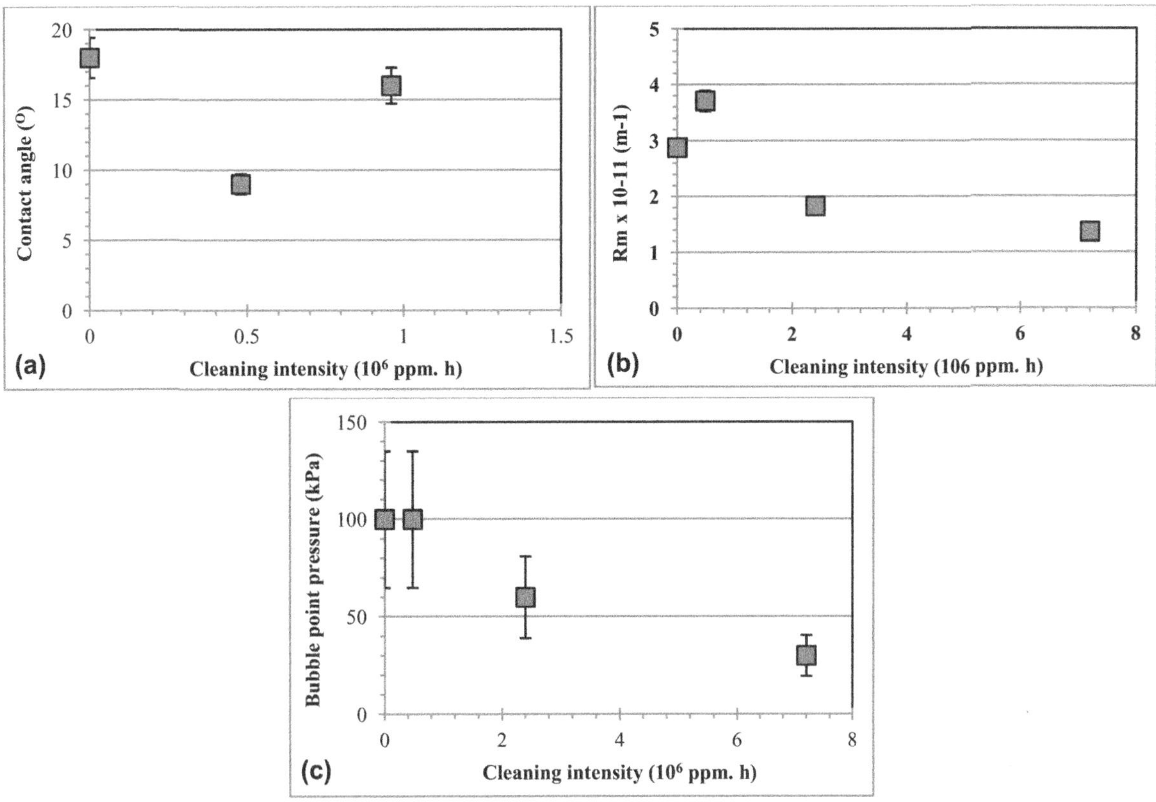

Figure 11- (a) Contact angle, (b) resistance and (c) bubble point of PES membrane aged with 2 % SDS.

Since this sudden increase and decrease in PES membrane resistance was not observed in the Bp pressure, these irregularities in the obtained trends can be due to accumulation of an excessive amount of SDS residue on the samples, which can explain the sharp decrease in the membrane contact angle and permeability. SDS was reported to cause fouling after membrane cleaning, as it can be difficult to remove from the membrane surface during rinsing time [23].

Similarly reports from the previous membrane samples, the decrease in the R_m values can be linked to the more hydrophilic membrane surface; whereas, the decreased Bp pressure was possibly due to pore opening of the negatively charged membrane.

3.4.3 Assessment of membrane degradation by HCl

Acid cleaning aims to remove inorganic scalants such as hardness and metal hydroxides. Mineral acids such as HCl are commonly used for PES and PVDF membranes. However, acid cleaning can lead to membrane damage in high concentrations due to the chlorination of the membrane. The alterations in the aged membrane samples and the possible degradation mechanisms involved are discussed in the following section.

3.4.3.1 PVDF-HF

According to Figure 12, PVDF-HF membrane became more hydrophilic after HCl ageing as contact angle decreased from 28 to 20 ^{0} for 0.96×10^4 ppm.h cleaning intensity. On the other hand, the membrane resistance increased slightly in the low HCl cleaning intensity. The membrane resistance decreased significantly from approximately 9 to 6×10^{11} m^{-1}, when 6×10^3 ppm.h cleaning intensity was applied.

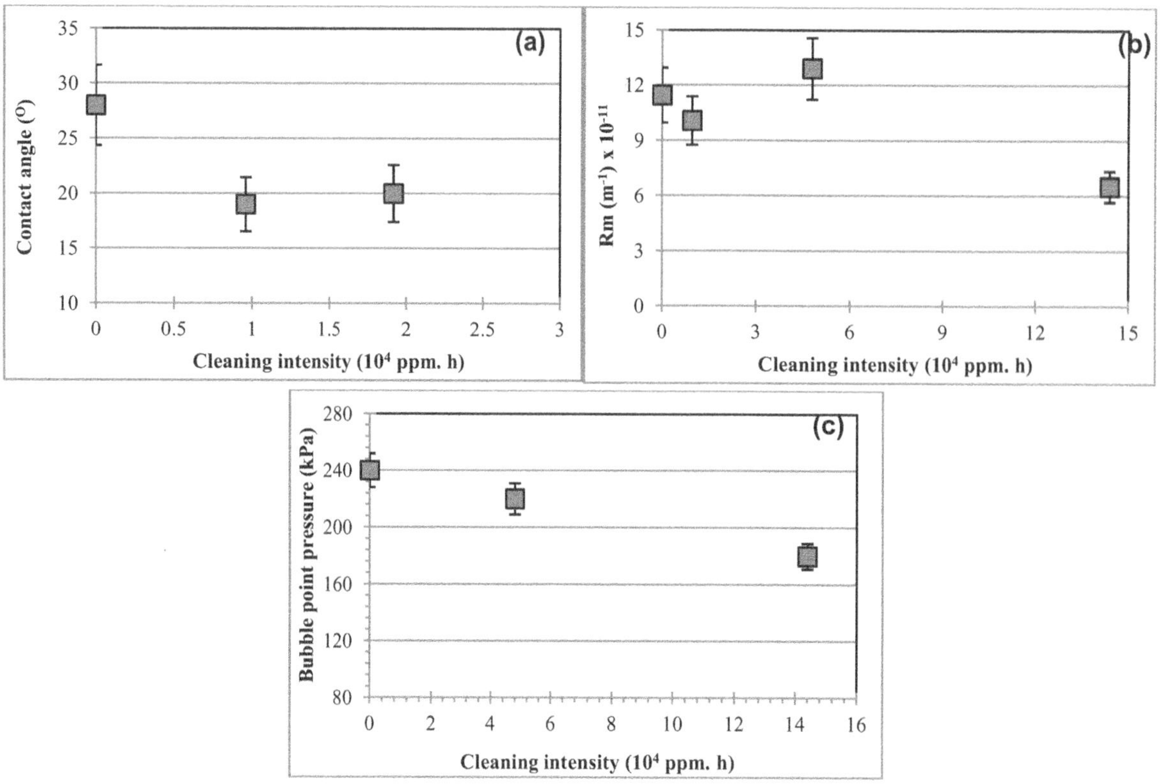

Figure 12- (a) Contact angle, (b) resistance and (c) bubble point of PVDF-HF membrane aged with 0.04% HCl.

In this study, the membrane became indeed more hydrophilic after HCl cleaning, as it was reported for polyethylene and polyamide membranes in previous studies [24-25]. The variation of Bp pressure for PVDF-HF membrane was observed to decrease gradually. In a previous study it was reported that acidic cleaning can led to a more negatively charged NF membrane, and opening of membrane pores due to swelling [22]. The decrease in the membrane resistance and Bp pressure was possibly due to membrane swelling, in the current study.

The FTIR spectrum of the HCl-aged PVDF-HF membrane (Appendix C) indicated a slight suppression in the carbonyl peaks at 1653 and 1734 cm^{-1}, which correspond to membrane additives. However, no significant alteration was observed in the rest of the peaks representing the PVDF material.

In a previous study, a significant decrease in water permeability for the aged sample was also observed after high pressure membranes were immersed in HCl solution [26]. The compaction of the polymer chains, associated to partial reduction of additives from the polyamide/polysulfone membrane structure, was related to the loss in membrane permeability. Other studies also reported the hydrophilizing effect of inorganic acids during membrane cleaning [24-25]. The contradiction between the results from the current study and the previous studies regrading membrane cleaning with HCl, indicate that acidic cleaning can lead to both increase and decrease in membrane surface charge depending on the membrane material used. Further studies are required to investigate the different factors affecting the membrane modification during acidic cleaning.

3.4.3.2 PVDF-HPI

The impact of HCl cleaning on the PVDF-HPI membrane was not significant, as can be observed (Figure 13). Although the trends in membrane contact angle suggested that the aged PVDF-HPI membrane surface was more hydrophilic (76^0), these changes did not exceed the SD values calculated for these measurements. On the other hand, the FTIR spectra of the aged PVDF-HPI membrane indicated that both membrane additive peaks, which were previously observed at 1650 and 1730 cm^{-1}, were suppressed.

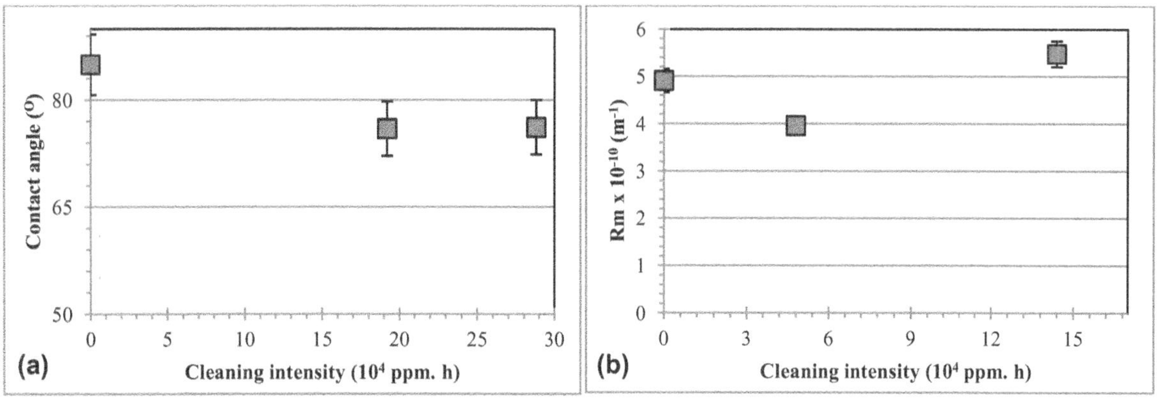

Figure 13- (a) Contact angle, (b) resistance and PVDF-HPI membrane aged with %0.04 HCl.

3.4.3.3 PVDF-HPO

The PVDF-HPO membrane aged with HCl behaved differently, compared to the other two PVDF membranes. As can be seen in Figure 14, the membrane contact angle and resistance showed contradictory trends after HCl ageing, although these changes were not significant. Therefore, this batch of PVDF-HPO aged in HCl samples was probably not reliable and needs further attention.

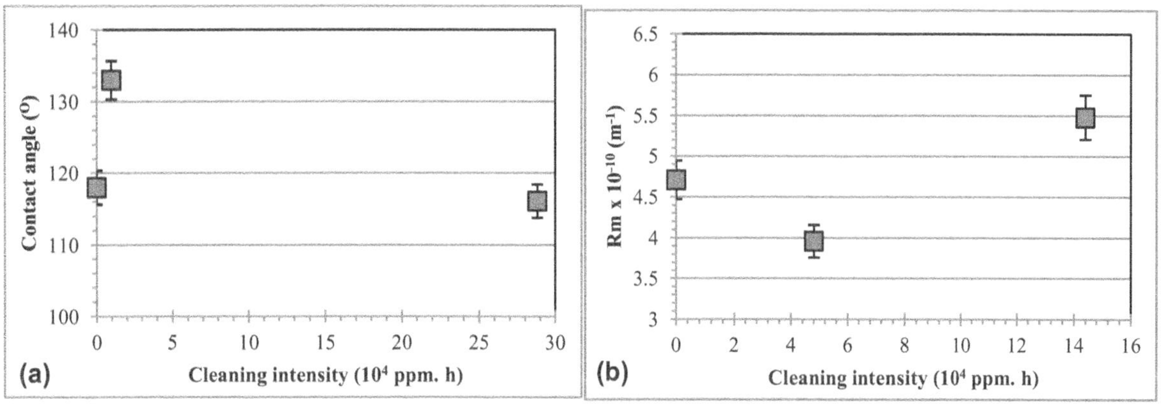

Figure 14- (a) Contact angle, (b) resistance PVDF-HPO membrane aged with %0.04 HCl.

The FTIR spectrum of the PVDF-HPO membrane (see Appendix A) showed new carbonyl peaks appearing after HCl ageing. The new carbonyl peaks in the aged membrane spectra can explain the increased membrane hydrophobicity. As explained previously, new carbonyl bands in the FTIR spectra can also indicate that membrane polymer skeleton degraded through cross linking between the PVDF chains. It is not worthy that the intensity of the new carbonyl peaks were rather small, which might be the reason which these cross linking was not observed in TGA and tensile test results.

3.4.3.4 PP

The trends obtained for aged PP membrane (Figure 15) indicated that HCl ageing resulted in a more hydrophilic membrane, as the membrane contact angle decreased from 28 to 12 ^{0}. The membrane resistance also decreased from 9.5 to $2.6x10^{11}$ m^{-1} after $6x10^{3}$ ppm.h of HCl ageing. The decrease in membrane contact angle and resistance confirmed the hydrophilising effect of HCl cleaning on PP membrane. However, the observed decrease was more gradual in terms of membrane resistance. These trends suggested that, although membrane hydrophilisation started at low HCl ageing intensity, its impact on the membrane resistance was only observed when $1.4x10^{5}$ ppm.h of HCl was applied.

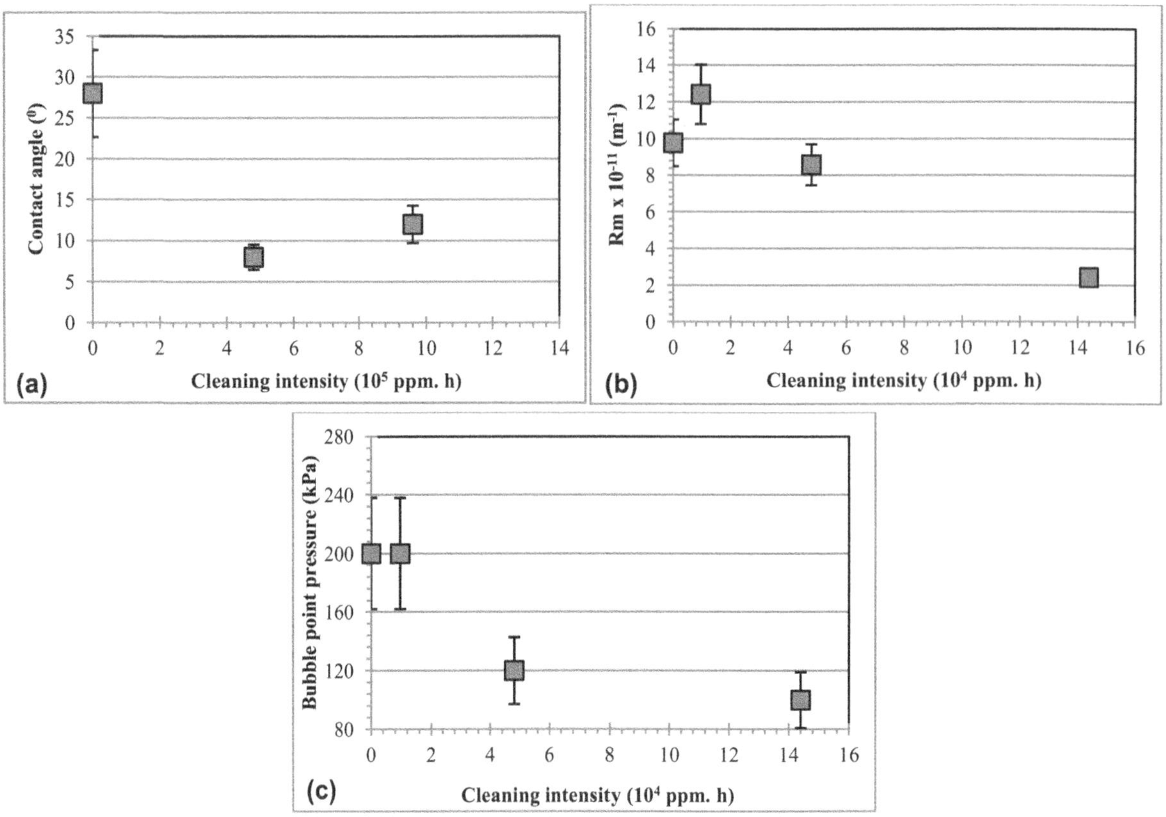

Figure 15- (a) Contact angle, (b) resistance and (c) bubble point of PP membrane aged with % 0.04 HCl.

The Bp pressure of the PP membrane decreased significantly at 9.6 × 10 ^{3} ppm.h of HCl cleaning. The FTIR spectrum (Appendix C) of the aged PP membrane by HCl at the maximum cleaning intensity did not show any significant alterations in the membrane polymer structure. Moreover, HCl did not result in the removal of the membrane additives. Therefore, the observed decrease in the membrane Bp pressure indicated that the membrane was probably due to pore opening and swelling of the more negatively charged membrane [22].

3.4.3.5 PES

Figure 16-(a-b) shows the properties of the aged PES membranes in HCl solution. As expected, a general small increase in hydrophilicity of the PES membrane was observed after HCl ageing. The observed decrease in the membrane resistance confirmed the occurrence of hydrophilisation of PES by HCl.

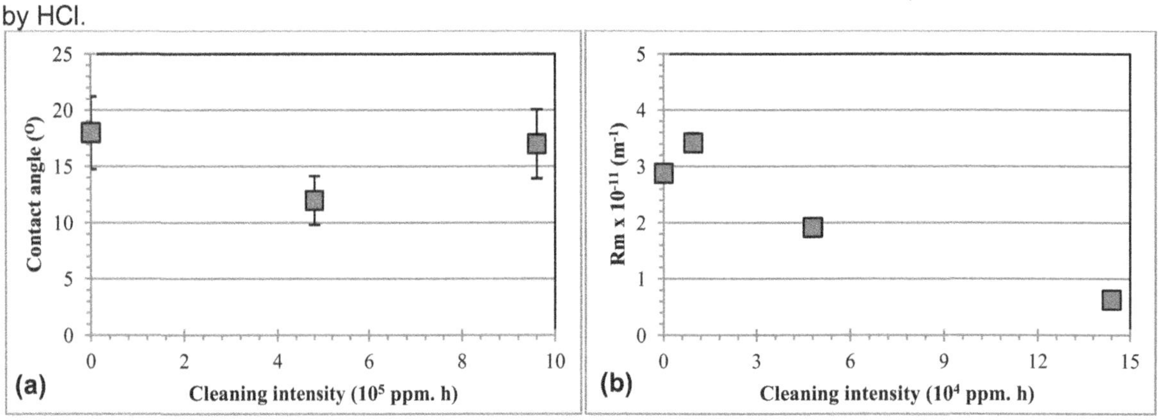

Figure 16- (a) Contact angle, (b) resistance of PES membrane aged with %0.04 HCl.

The PES membrane degradation due to HCl cleaning was not as significant as the other hollow fibre membrane tested in this study. For example, based on Bp pressure results, PES membrane pore size was not compromised (Table 14). The FTIR spectra of the aged PES membrane (Appendix C) confirmed that the HCl did not significantly change the chemical structure of the PES membrane.

3.4.4 Assessment of membrane degradation by NaOH

Caustic cleaners (up to pH 12 depending on the membrane) such as NaOH are particularly effective for removal of organics and proteins. These chemicals act by breaking bonds between the membrane surface and the fouling material through hydrolysis or oxidation and thus, solubilise the foulants.

3.4.4.1 PVDF_HF

The PVDF-HF membrane properties were significantly compromised after ageing with NaOH (Figure 17). Following the minimum intensity of NaOH ageing, the membrane contact angle, resistance, Bp pressure and maximum elongation were all observed to significantly decrease. For example, the membrane contact angle decreased from 28 to 19 °.

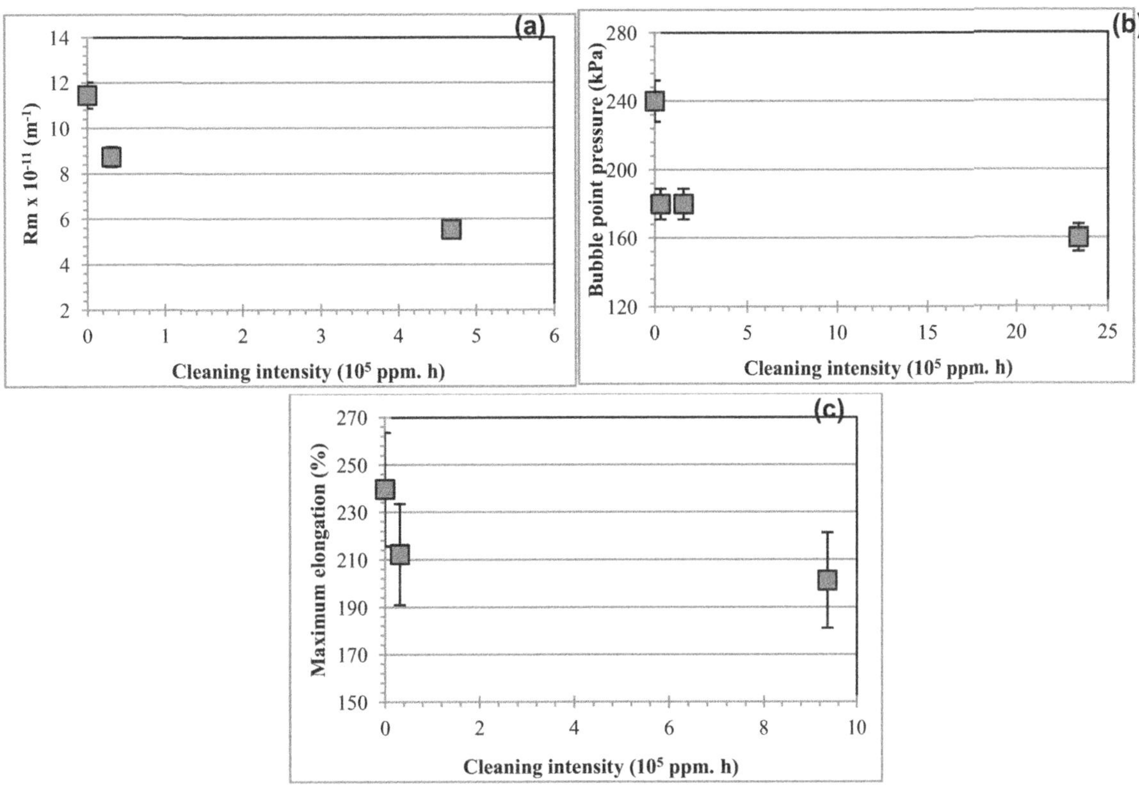

Figure 17- (a) Resistance, (b) bubble point pressure and (c) maximum elongation of PVDF-HF membrane aged with %0.13 NaOH.

The FTIR spectra of the NaOH aged PVDF-HF membrane (see Appendix D) showed that the carbonyl peak at 1734 cm^{-1}, which most probably corresponds to membrane additive, was removed completely at 9.6x10^{5} ppm.h NaOH ageing. The vulnerability of PVDF material to NaOH exposure has been reported previously [27-28]. The chemical attack of NaOH to the PVDF-HF was even observed through the membrane colour change from white to brown. The colour change in NaOH-aged PVDF membrane was reported to be due to dehydrofluorination phenomena, in which the C=C are formed as a result of the elimination of H-F units from the polymer [27]. However, the only significant change in the aged membrane FTIR spectra (Appendix D) was the removal of the carbonyl peak at 1735 cm^{-1}. Although the FTIR analysis did not show any new C=C bands forming in the aged membrane spectra, the membrane colour change can confirm the occurrence of dehydrofluorination phenomena.

3.4.4.2 PVDF-HPI

The properties of PVDF-HPI membrane also indicated that the NaOH exposure resulted in the degradation of the membrane. However, in comparison with the PVDF-HF membrane, the loss of integrity for the flat sheet samples was less pronounced.

The membrane became significantly more hydrophilic (77 ^{0}) as the NaOH cleaning intensity increased to 9.36x10^5 ppm.h. On the other hand, membrane resistance increased significantly (from 4.91 to 6.52 × 10^{10} m^{-1}), whereas the membrane maximum elongation decreased slightly to 124 %. The membrane embrittlement can result from the occurrence of dehydrofluorination of the PVDF polymer, as mentioned previously. Nevertheless, the colour change due to NaOH exposure was only observed for the PVDF-HF membrane.

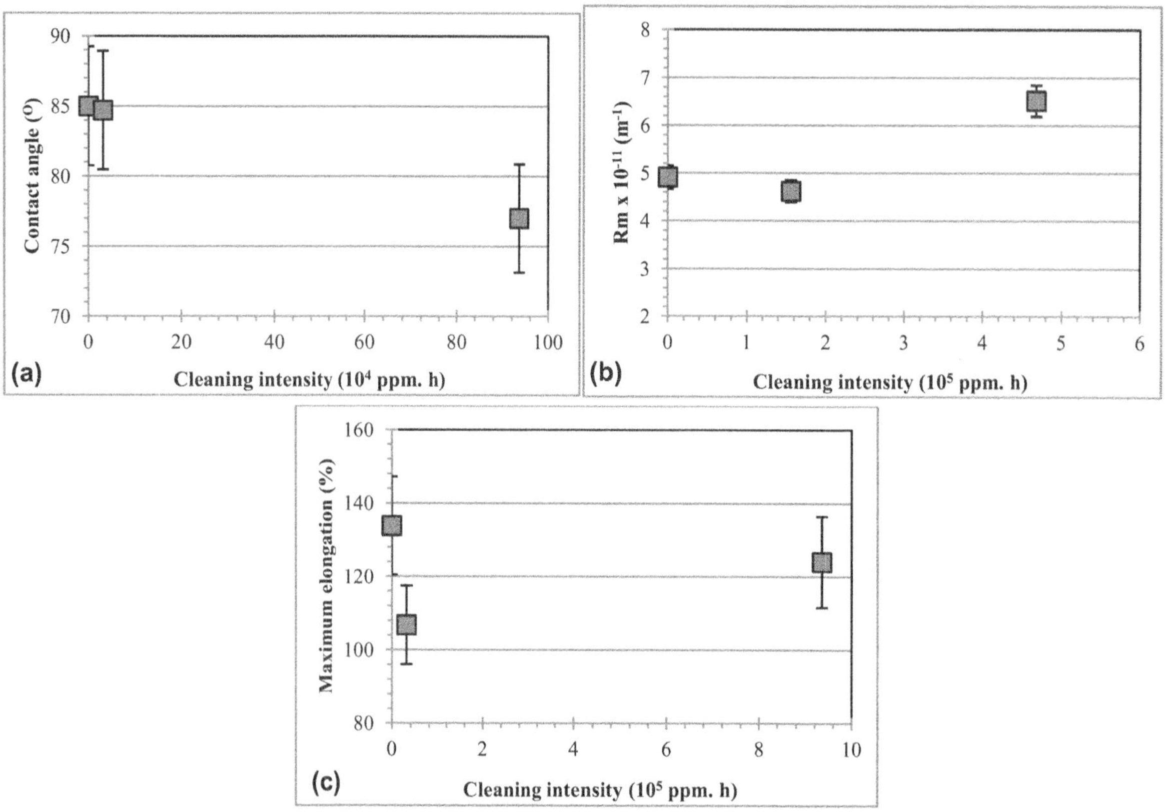

Figure 18- (a) Contact angle, (b) resistance, (c) maximum elongation of PVDF-HPI membrane aged with %0.13 NaOH.

Finally, the FTIR spectra of the aged PVDF-HPI membrane (Appendix D) indicated that the carbonyl peaks, corresponding to the additives, were whether suppressed or completely removed as the result of NaOH exposure.

3.4.4.3 PVDF-HPO

The changes in the PVDF-HPO membrane properties were not significant in terms of its R_m, CA and tensile properties (Figure 19). The decrease in the membrane maximum elongation indicated slight signs of embrittlement in the aged membrane after NaOH exposure.

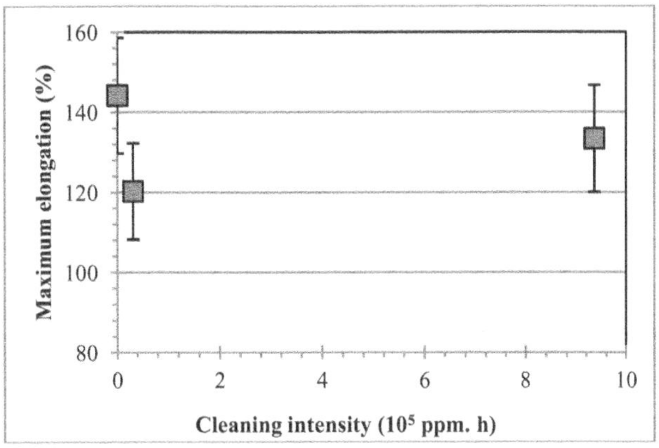

Figure 19- Maximum elongation of PVDF-HPO membrane aged with %0.13 NaOH.

The FTIR spectra of the NaOH-aged PVDF-HPO membrane (Appendix D) indicated that a new carbonyl peak at 1650 cm^{-1} emerged. This indicates that membrane polymer was degraded through cross linking.

3.4.4.4 PP

The alterations in PP membrane properties aged in NaOH solution was similar to those observed with PVDF-HF, although the extent of damage to PP was less significant. For example, NaOH treatment did not result in colour change and embrittlement of the PP membrane. The significant alterations in PP properties involved decrease in the membrane contact angle from 28 to 14 ^{0} at 3.12×10^4 ppm.h of NaOH ageing (Table 7). The aged PP membrane resistance also decreased from approximately 10 to 2×10^{11} m^{-1} at the 1.95×10^4 ppm.h of NaOH cleaning. Moreover, the membrane Bp pressure decreased to less than half of its initial value due to NaOH exposure indicating that the PP membrane pore size was severely compromised.

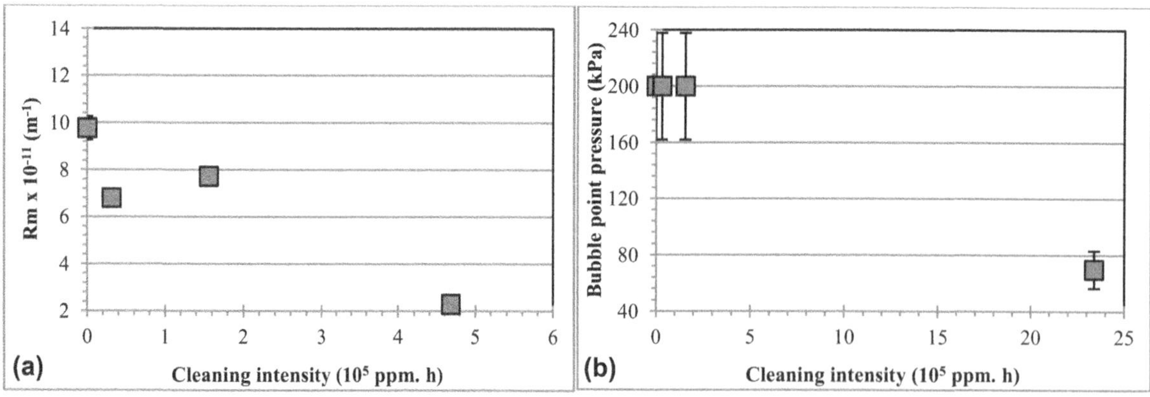

Figure 20- (a) Resistance and (b) bubble point pressure of PP membrane aged with 0.13 % NaOH.

It was previously reported that PP polymer in contact with sodium alkaline solutions may convert the carboxylic groups to carboxylates as the result of oxidative degradation [29] which can produce a new infrared absorption and shift from 1720cm^{-1} wave number. The new carboxylate absorption band corresponding to the sodium carboxylate (–COONa) has been reported to appear between 1545 and 1580cm^{-1} [29]. However, the FTIR spectrum of the NaOH-aged PP membrane (see Appendix D) only difference from the wetted sample was an emerging hydroxyl peak at 3640 cm^{-1}. Those stringent alterations reported for PP polymer in the later study, was probably observed at higher temperature and/or concentrations of NaOH.

3.4.4.5 PES

The TGA analysis of the PES aged membrane by NaOH in Figure 21-(a) at the maximum cleaning intensity showed that the T_{max} of the membrane shifted from 526 to 485 °C. This indicated that the PES membrane polymer reaction with NaOH has resulted in chain scission of the polymer. The PES membrane contact angle declined significantly at the minimum NaOH cleaning intensity, as can be seen in Figure 21 (b). The membrane resistance and Bp pressure also generally decreased however, a significant increase in both properties was observed at the minimum NaOH cleaning intensity (0.312x10^5 ppm.h). As the NaOH dose increased, the membrane resistance and Bp pressure suddenly decreased. The decreased Bp pressure indicated that the PES membrane pore size was severely compromised.

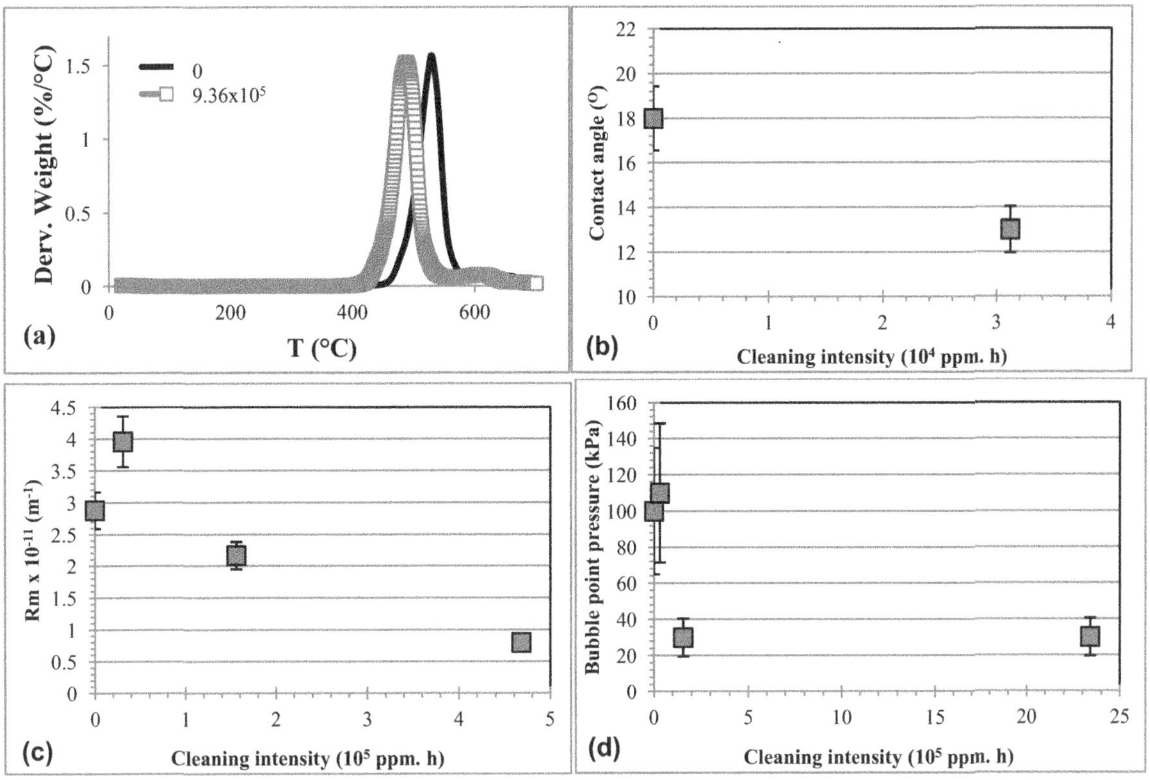

Figure 21- (a) Contact angle, (b) resistance, (c) bubble point pressure and (d) maximum elongation of PES membrane aged with %0.13 NaOH.

The FTIR spectra of PES membrane aged with NaOH indicated that the peak at 1500 cm^{-1}, corresponding to C-C functional groups, was suppressed. These results indicated that PES material suffered from NaOH attack (hydrolysis) which led to chain scission. The FTIR results also confirmed the trend observed by TGA.

4. CONCLUSIONS

Four different cleaning agents (NaOCl, SDS, HCl and NaOH) were used to age five commercial porous membranes fabricated from various materials. The membranes were in both hollow fibre (PVDF-HF, PP and PES) and flat sheet (PVDF-HPI and –HPO) configurations. The maximum duration of the ageing experiment was 30 days, and membrane samples were analysed for a wide range of, cleaning intensities. Following these ageing experiments, a range of selected analytical techniques were carried out, allowing a detailed assessment of the membrane characteristics (both physical and chemical).

The results obtained from the membrane characterisation allowed (1) an arbitrary ranking of membrane stability against different chemical cleaning agents (2) a better understanding of the mechanisms occurring during membrane ageing. The membrane ranking indicated that PVDF-HF membrane properties were significantly modified after NaOCl and NaOH, whereas, SDS-aged membranes were not significantly affected. Both PVDF flat sheet membranes (PVDF-HPI and PVDF-HPO) were more sensitive to HCl exposure. Modifications to PP membrane properties were equally observed after NaOCl and HCl ageing. PES membrane was observed to be more sensitive to SDS exposure and, seemed to be less compromised after NaOH ageing.

Further assessment of the various membrane samples modification/degradation mechanisms by different cleaning agents are summarised in Table 18. All oxidising agents (NaOCl, HCl and NaOH) generally resulted in removal of membrane additives. For example, PVDF-HF membrane additives were removed by all three oxidising agents, although the extent of degradation of additives varied for different cleaning chemicals. NaOCl attack generally resulted in loss of membrane integrity through cross linking in PVDF, chain scission in PP, and the chlorination of PES.

Table 18- Summary of the proposed degradation mechanisms by various cleaning agents for different membranes.

		SDS	HCl	NaOH
PVDF-HF	• Elimination of additives • Cross linking followed by chain scission	• Hydrophilisation	• Elimination of additives • Hydrophilisation	• Elimination of additives • Dehydrofluorination
PVDF-HPI	• Elimination of additives • Cross linking followed by chain scission	• Elimination of additives	• Elimination of additives • Hydrophilisation	• Elimination of additives • Dehydrofluorination
PVDF-HPO	• Cross linking	• More hydrophobic	-	• Cross linking
PP	• Elimination of additives • Chain scission	• Hydrophilisation • Swelling	• Hydrophilisation • Pore enhancement	• Hydrolysis • Chain scission
PES	• Chlorination of C-S bonds	• Hydrophilisation • Swelling	• Hydrophilisation	• Hydrolysis • Chain scission

"-": Not available.

SDS cleaning led to subtle changes in membrane surface characteristics due to attachment of its hydrophobic tail on the membrane surface, which resulted in more hydrophilic and negatively charged membrane. As the result of membrane swelling, the membrane apparent pore size also seemed to increase. The tested membranes exposure to HCl only resulted in modification of membrane surface properties (hydrophilisation), despite removal of additives. Therefore, HCl seemed to be the less aggressive chemical amongst the three oxidising agents used in this study possibly due to the low concentration applied. NaOCl and NaOH both resulted in degradation of all membranes core material.

Similar to NaOCl, NaOH resulted in significant chemical degradation of the tested membranes. NaOH exposure generally resulted in membrane degradation through mostly dehydrofluorination in PVDF and chain scission through hydrolysis in PP and PES.

Finally, the relative contradiction observed between the testing methods indicated their limitations in measuring a single parameter. For example, many of the membrane characteristics (such as pore size, charge and roughness) can affect the contact angle value obtained from the goniometer. On the other hand, trends obtained from contact angle, membrane resistance and bubble point tests were successfully used to validate some proposed hypothesis regarding degradation mechanisms.

5. REFERENCES

[1]. Causserand, C., S. Rouaix, J.-P. Lafaille, and P. Aimar, Ageing of polysulfone membranes in contact with bleach solution: Role of radical oxidation and of some dissolved metal ions, Chemical Engineering and Processing: Process Intensification. 47 (2008) 48-56.

[2]. Gaudichet-Maurin, E. and F. Thominette, Ageing of polysulfone ultrafiltration membranes in contact with bleach solutions, Journal of Membrane Science. 282 (2006) 198-204.

[3]. Puspitasari, V., A. Granville, P. Le-Clech, and V. Chen, Optimum Cleaning and Ageing Effect of Sodium Hypochlorite on Polyvinylidene Fluoride (PVDF) Membrane, (2010).

[4]. Zondervan, E., A. Zwijnenburg, and B. Roffel, Statistical analysis of data from accelerated ageing tests of PES UF membranes, Journal of Membrane Science. 300 (2007) 111-116.

[5]. Trägårdh, G., Membrane Cleaning, Desalination. 71 (1989) 325-335.

[6]. Shorrock, C.J. and M.R. Bird, Membrane Cleaning: Chemically Enhanced Removal of Deposits Formed During Yeast Cell Harvesting, Food and Bioproducts Processing. 76 (1998) 30-38.

[7]. Pall Corporation. *Membrane Chemical Cleaning: From Art to Science*. 2006 [cited 2009 5th July]; Available from: http://www.pall.com/pdf/mtcpaper.pdf.

[8]. Pearce, G., Introduction to membranes: Membrane selection, Filtration & Separation. 44 (2007) 35-37.

[9]. Bégoin, L., M. Rabiller-Baudry, B. Chaufer, M.-C. Hautbois, and T. Doneva, Ageing of PES industrial spiral-wound membranes in acid whey ultrafiltration, Desalination. 192 (2006) 25-39.

[10]. Arkhangelsky, E., U. Goren, and V. Gitis, Retention of organic matter by cellulose acetate membranes cleaned with hypochlorite, Desalination. 223 (2008) 97-105.

[11]. Kang, G.-D., C.-J. Gao, W.-D. Chen, X.-M. Jie, Y.-M. Cao, and Q. Yuan, Study on hypochlorite degradation of aromatic polyamide reverse osmosis membrane, Journal of Membrane Science. 300 (2007) 165-171.

[12]. Yadav, K., K. Morison, and M.P. Staiger, Effects of hypochlorite treatment on the surface morphology and mechanical properties of polyethersulfone ultrafiltration membranes, Polymer Degradation and Stability. 94 (2009) 199501961.

[13]. Peniche, C., D. Zaldívar, M. Pazos, S. Páz, A. Bulay, and J.S. Román, Study of the thermal degradation of poly(<I>N</I>-vinyl-2-pyrrolidone) by thermogravimetry-FTIR, Journal of Applied Polymer Science. 50 (1993) 485-493.

[14]. Mark, J. and Knovel, Polymer data handbook, Oxford University Press New York, 1999.

[15]. Fayolle, B., L. Audouin, and J. Verdu, Oxidation induced embrittlement in polypropylene -- a tensile testing study, Polymer Degradation and Stability. 70 (2000) 333-340.

[16]. Gryta, M., J. Grzechulska-Damszel, A. Markowska, and K. Karakulski, The influence of polypropylene degradation on the membrane wettability during membrane distillation, Journal of Membrane Science. 326 (2009) 493-502.

[17]. Lazzari, M. and O. Chiantore, Thermal-ageing of paraloid acrylic protective polymers, Polymer. 41 (2000) 6447-6455.

[18]. Jacangelo, J.G. and R. K.Noack, System concepts.in Microfiltlration and ultrafiltration membranes for drinking water, American Water Works Association, Colorado, 2005. pp. 51-64.

[19]. Cheryan, M., Ultrafiltration and microfiltration handbook, Technomic Pub. Co., Lancaster, Pa., 1998.

[20]. Arkhangelsky, E., D. Kuzmenko, and V. Gitis, Impact of chemical cleaning on properties and functioning of polyethersulfone membranes, Journal of Membrane Science. 305 (2007) 176-184.

[21]. Al-Amoudi, A. and R.W. Lovitt, Fouling strategies and the cleaning system of NF membranes and factors affecting cleaning efficiency, Journal of Membrane Science. 303 (2007) 4-28.

[22]. Al-Amoudi, A., P. Williams, S. Mandale, and R.W. Lovitt, Cleaning results of new and fouled nanofiltration membrane characterised by zeta potential and permeability, Separation and Purification Technology. 54 (2007) 234-240.

[23]. Li, H. and V. Chen, Membrane Fouling and Cleaning in Food and Bioprocessing, Z.F. Cui and H.S. Muralidhara, Editors.in Membrane Technology, Butterworth-Heinemann, Oxford, 2010. pp. 213-254.

[24]. Dimov, A. and M.A. Islam, Hydrophilization of polyethylene membranes, Journal of Membrane Science. 50 (1990) 97-100.

[25]. Mukherjee, D., A. Kulkarni, and W.N. Gill, Chemical treatment for improved performance of reverse osmosis membranes, Desalination. 104 (1996) 239-249.

[26]. Benavente, J., M.I. Vázquez, and R. de Lara, Modification of active and porous sublayers of aged polyamide/polysulfone composite membranes due to HNO3 treatment: Effect of treatment time, Journal of Colloid and Interface Science. 297 (2006) 226-234.

[27]. Awanis Hashim, N., Y. Liu, and K. Li, Stability of PVDF hollow fibre membranes in sodium hydroxide aqueous solution, Chemical Engineering Science. In Press, Accepted Manuscript.

[28]. Marchand-Brynaert, J., N. Jongen, and J.L. Dewez, Surface hydroxylation of poly (vinylidene fluoride)(PVDF) film, Journal of Polymer Science Part A: Polymer Chemistry. 35 (1997) 1227-1235.

[29]. Scheirs, J., Compositional and failure analysis of polymers: a practical approach, John Wiley & Sons Inc, 2000.

6. APPENDICES

Appendix A: FTIR analysis of NaOCl-aged membrane

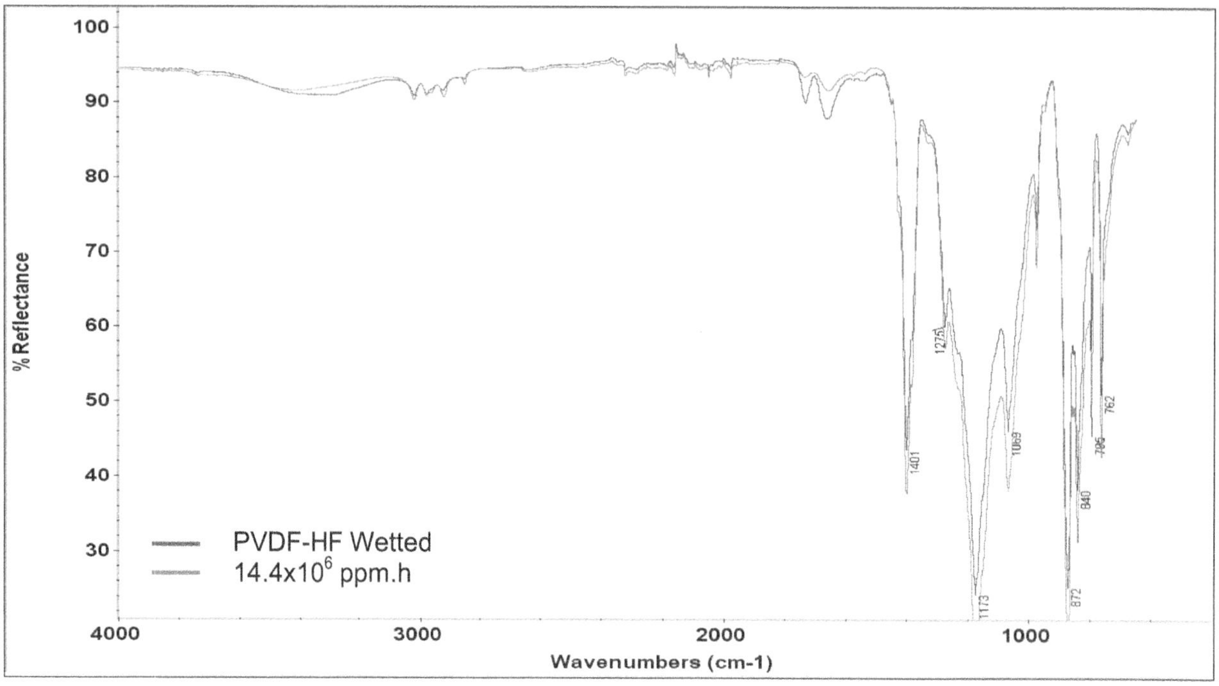

Figure A-1: FTIR spectrum of wetted and aged PVDF-HF membrane in %2 NaOCl.

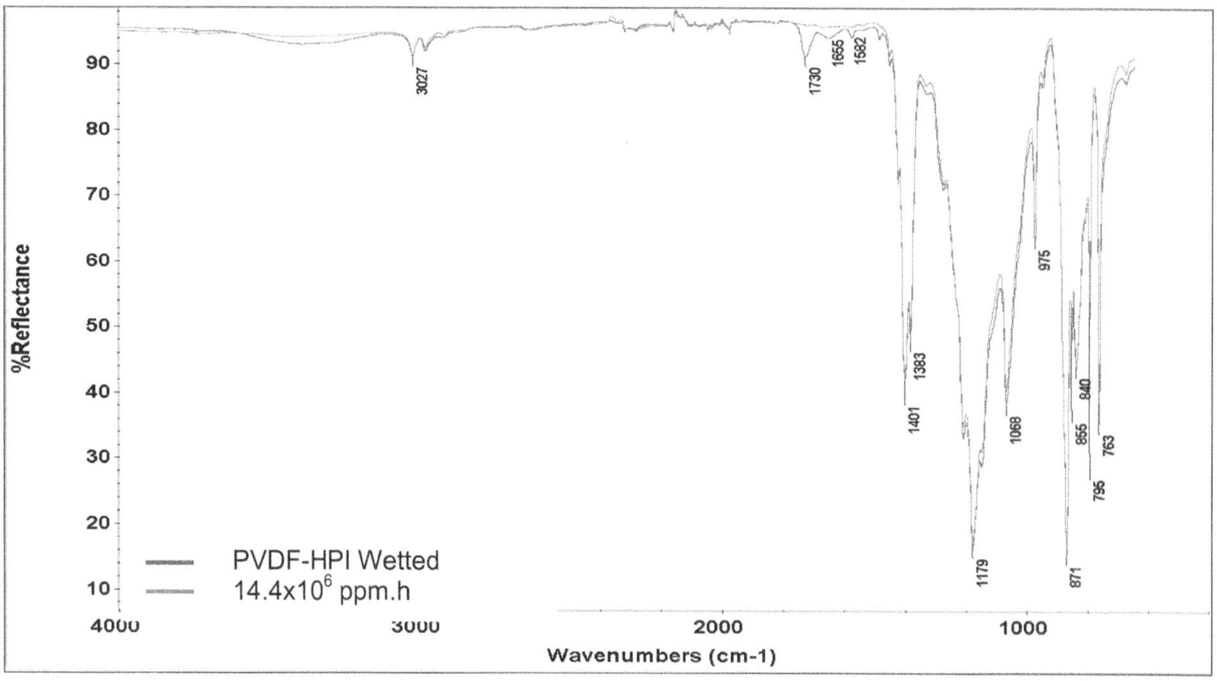

Figure A- 3: FTIR spectrum of wetted and aged PVDF-HPI membrane in % 2 NaOCl.

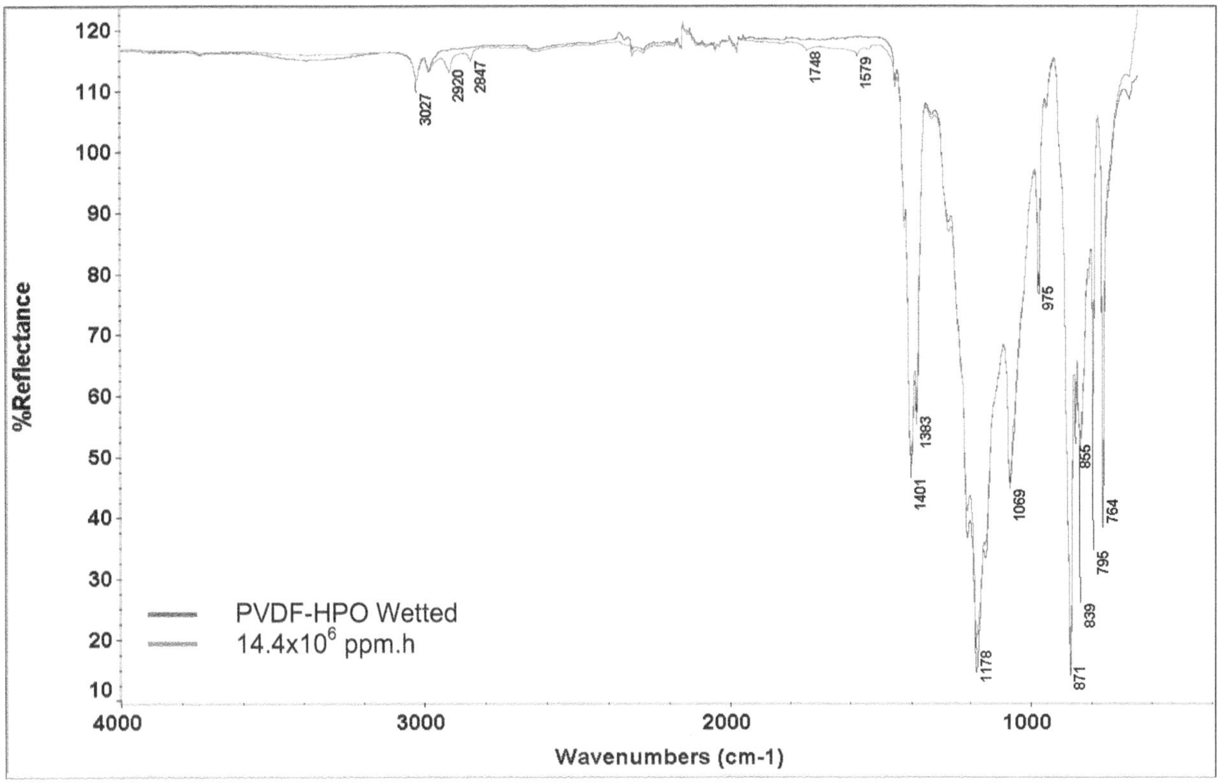

Figure A- 3: FTIR spectrum of wetted and aged PVDF-HPO membrane in 2 %NaOCl.

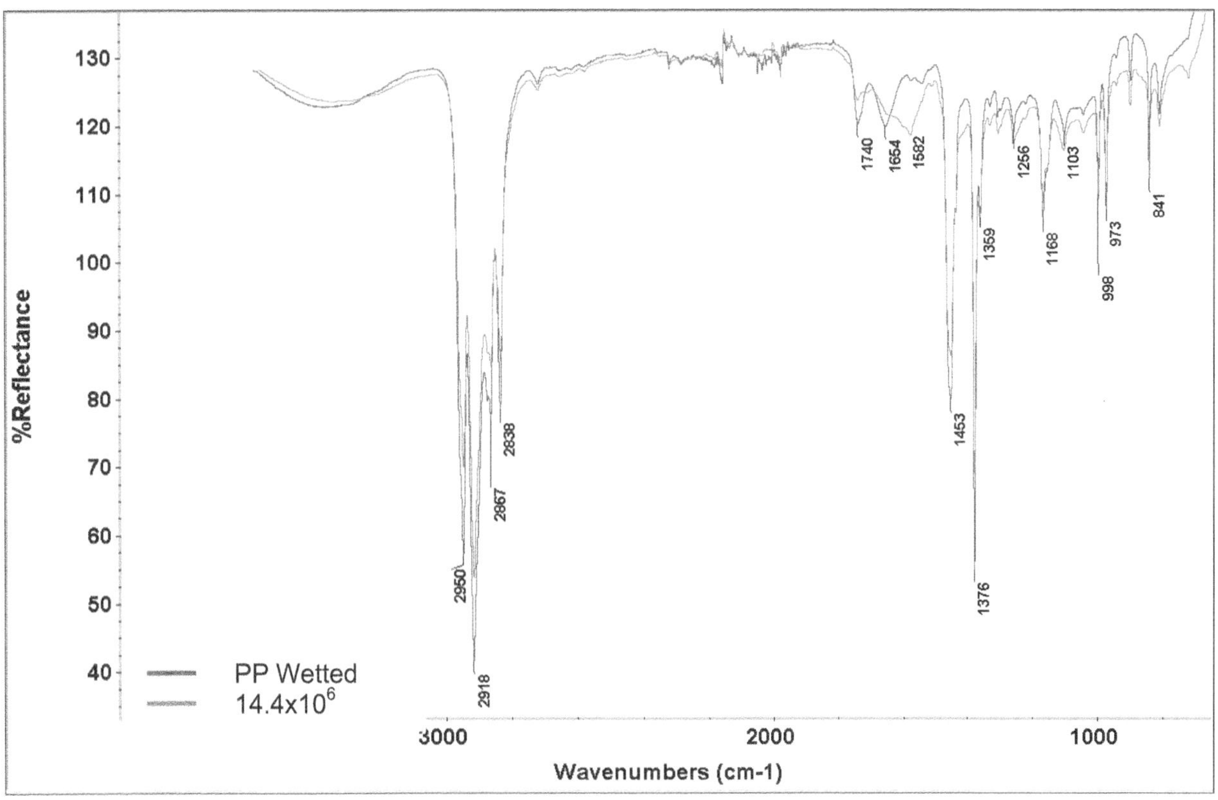

Figure A- 2: FTIR spectrum of wetted and aged PP membrane in 2% NaOCl.

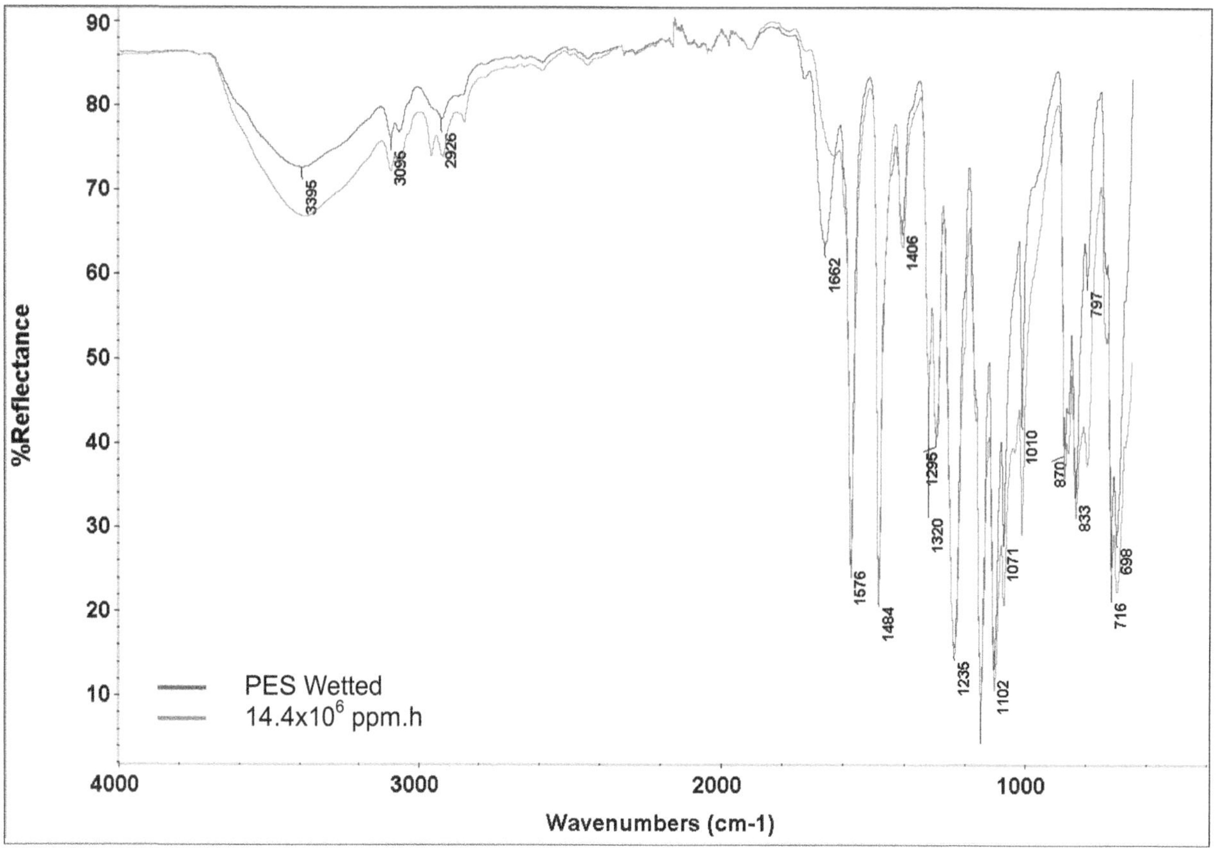

Figure A- 3: FTIR spectrum of wetted and aged PES membrane in 2% NaOCl.

Appendix B: FTIR analysis of SDS-aged membrane

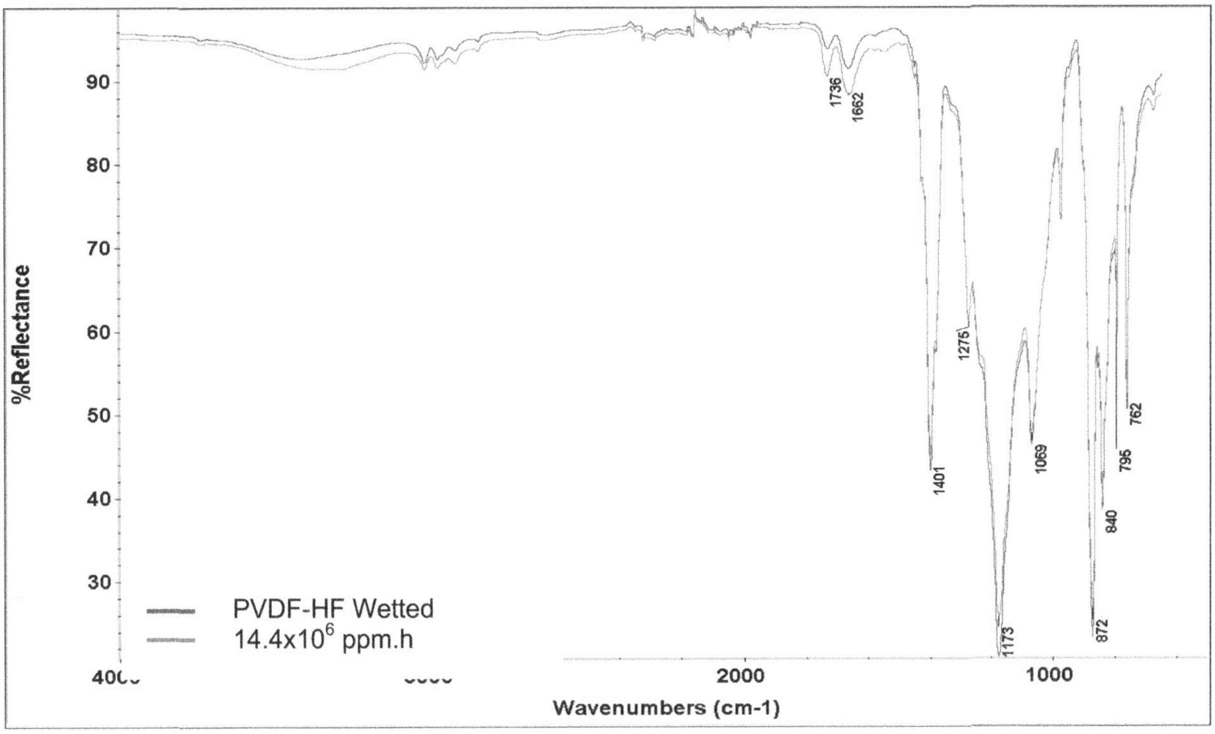

Figure B: FTIR spectrum of wetted and aged PVDF-HF membranes in 2% SDS.

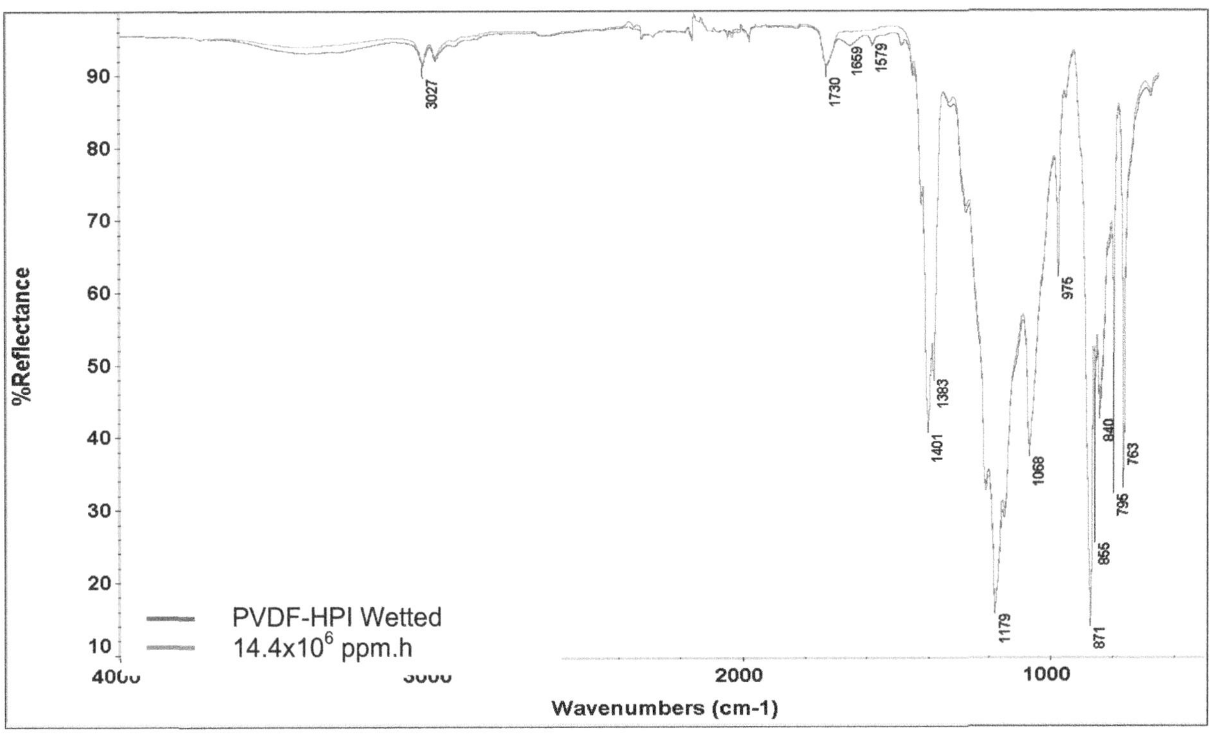

Figure B: FTIR spectrum of wetted and aged PVDF-HPI membranes in 2% SDS.

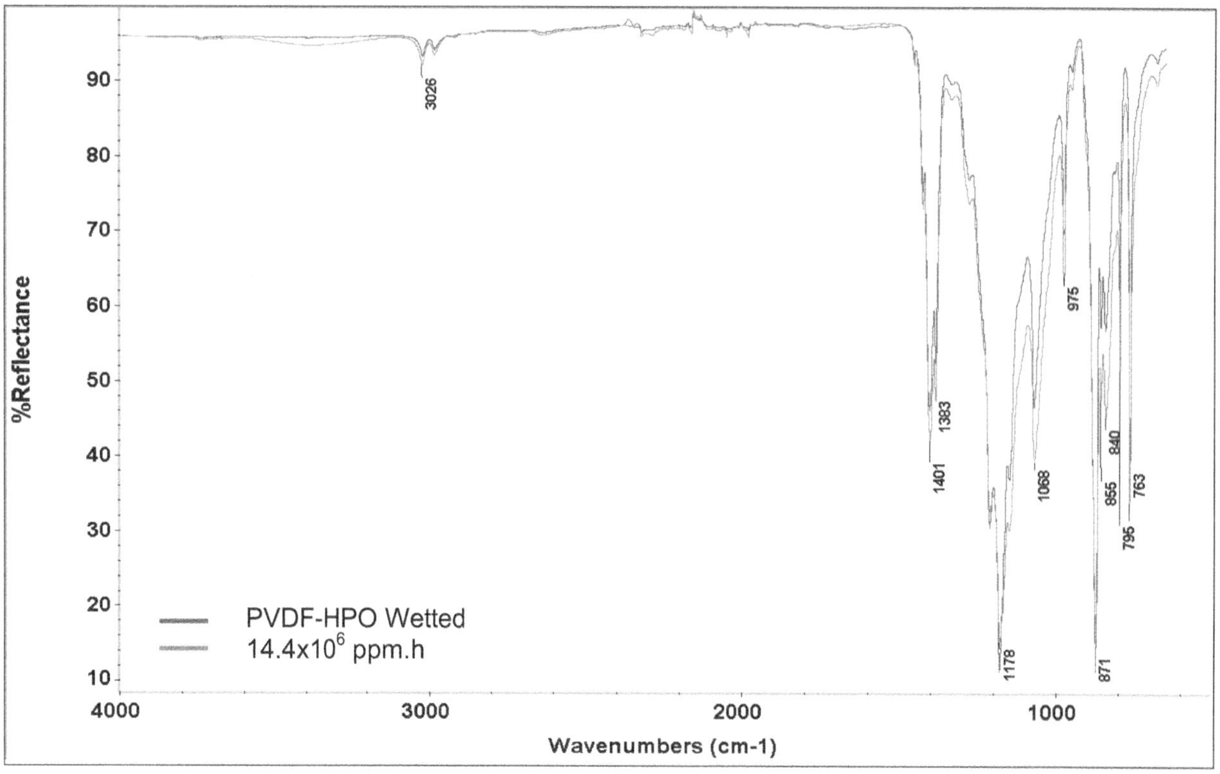

Figure B: FTIR spectrum of wetted and aged PVDF-HPO membranes in 2% SDS.

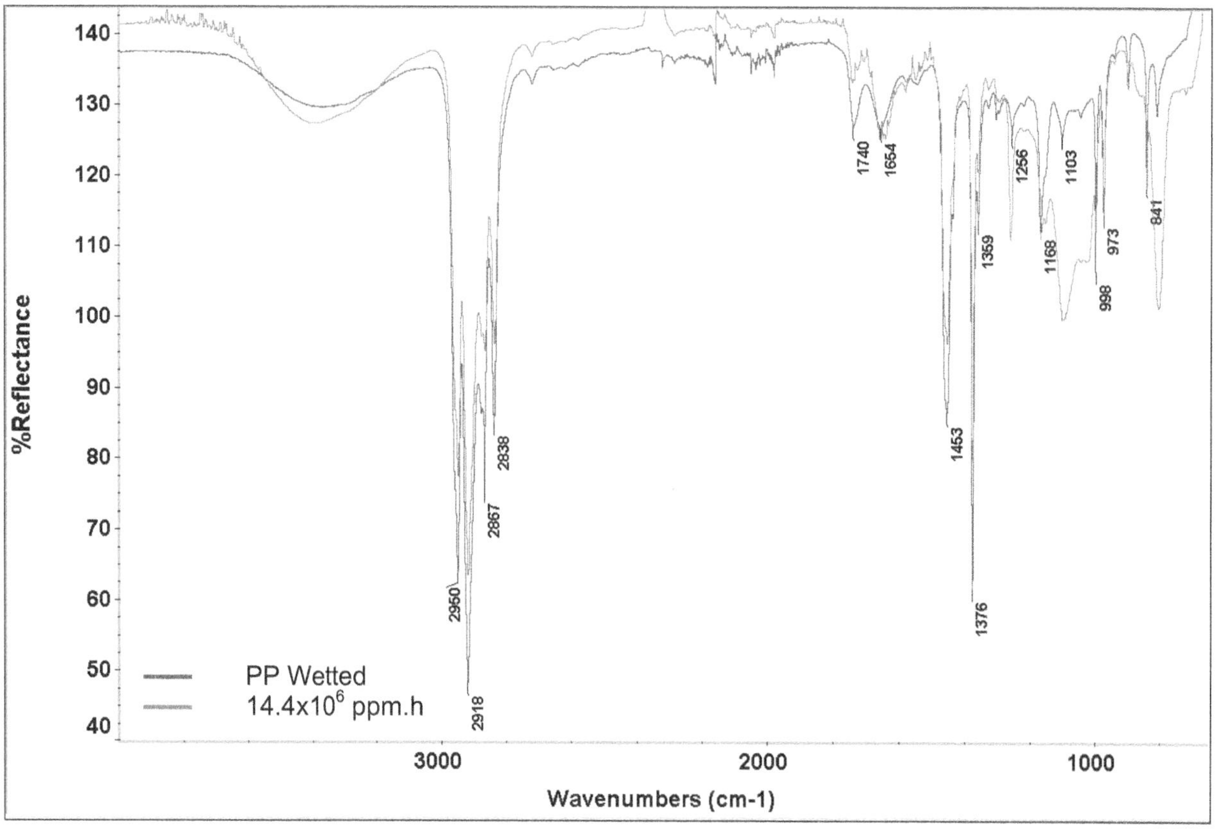

Figure B: FTIR spectrum of wetted and aged PP membranes in 2% SDS.

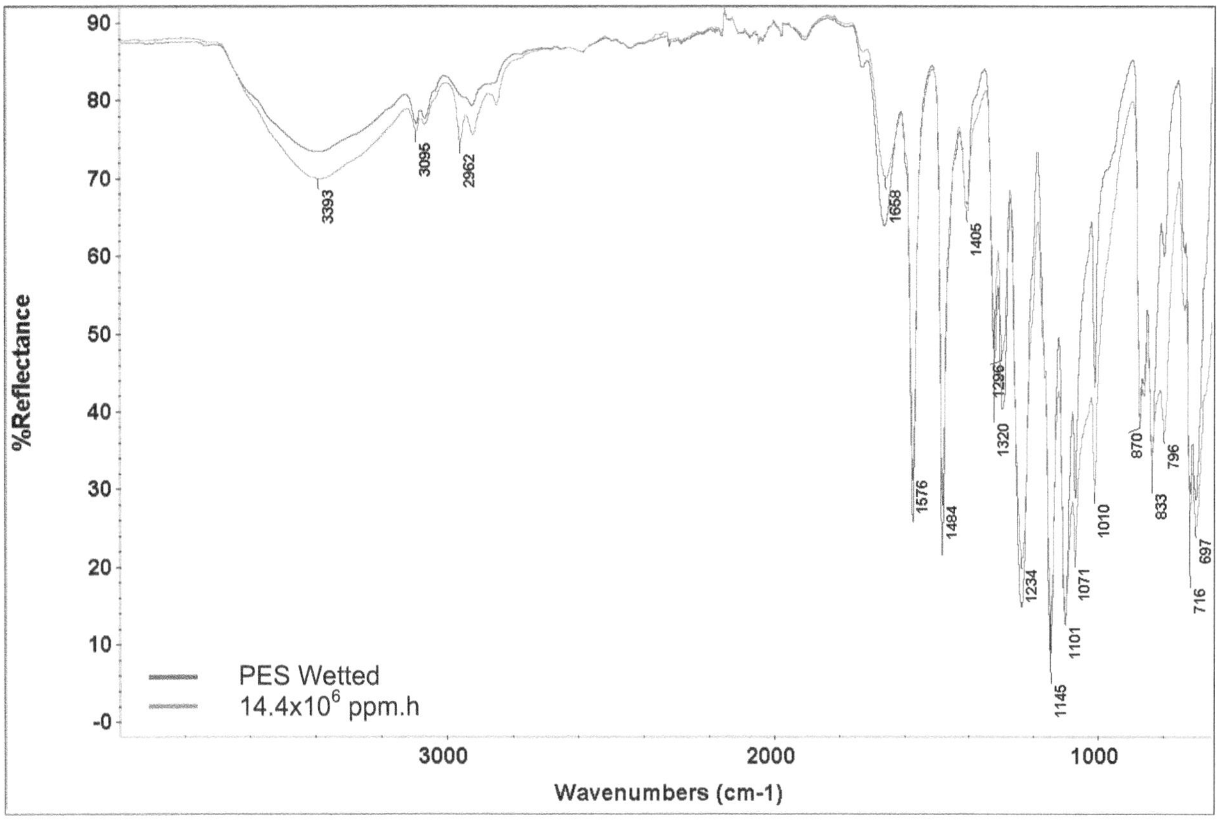

Figure B: FTIR spectrum of wetted and aged PES membranes in 2% SDS.

Appendix C: FTIR analysis of HCl-aged membrane

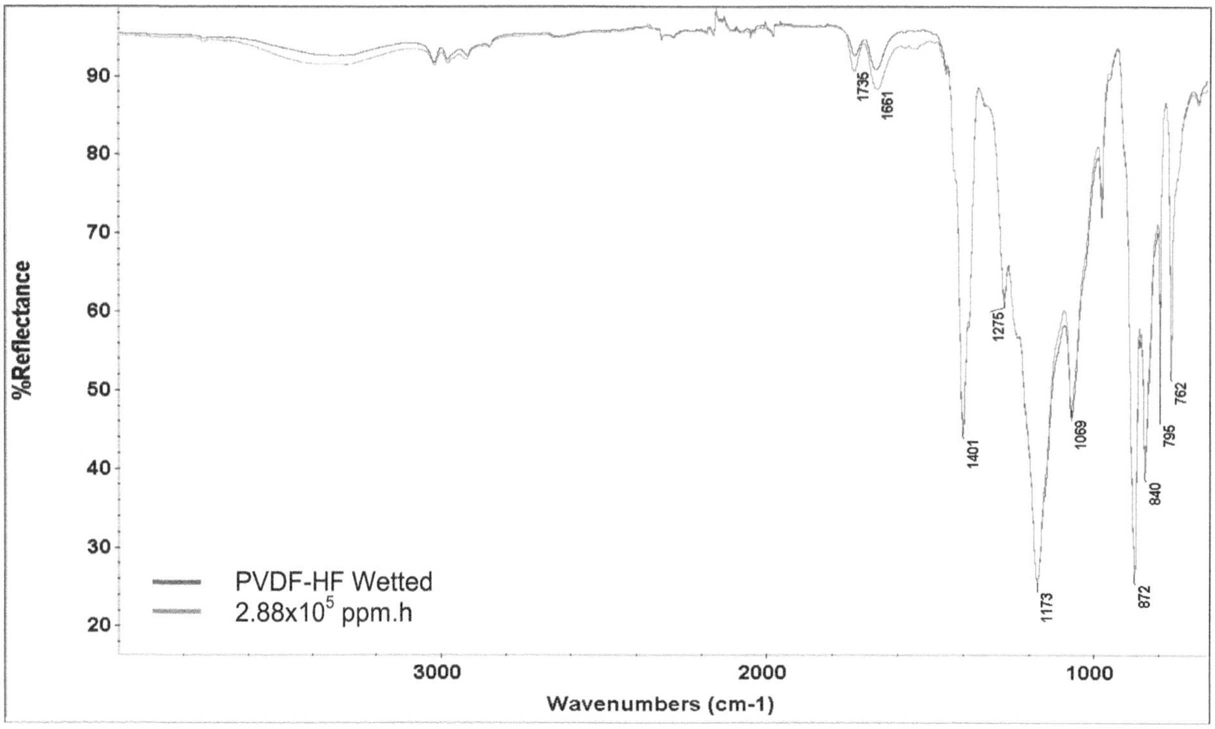

Figure C: FTIR spectrum of wetted and aged PVDF-HF membranes in 0.04% HCl.

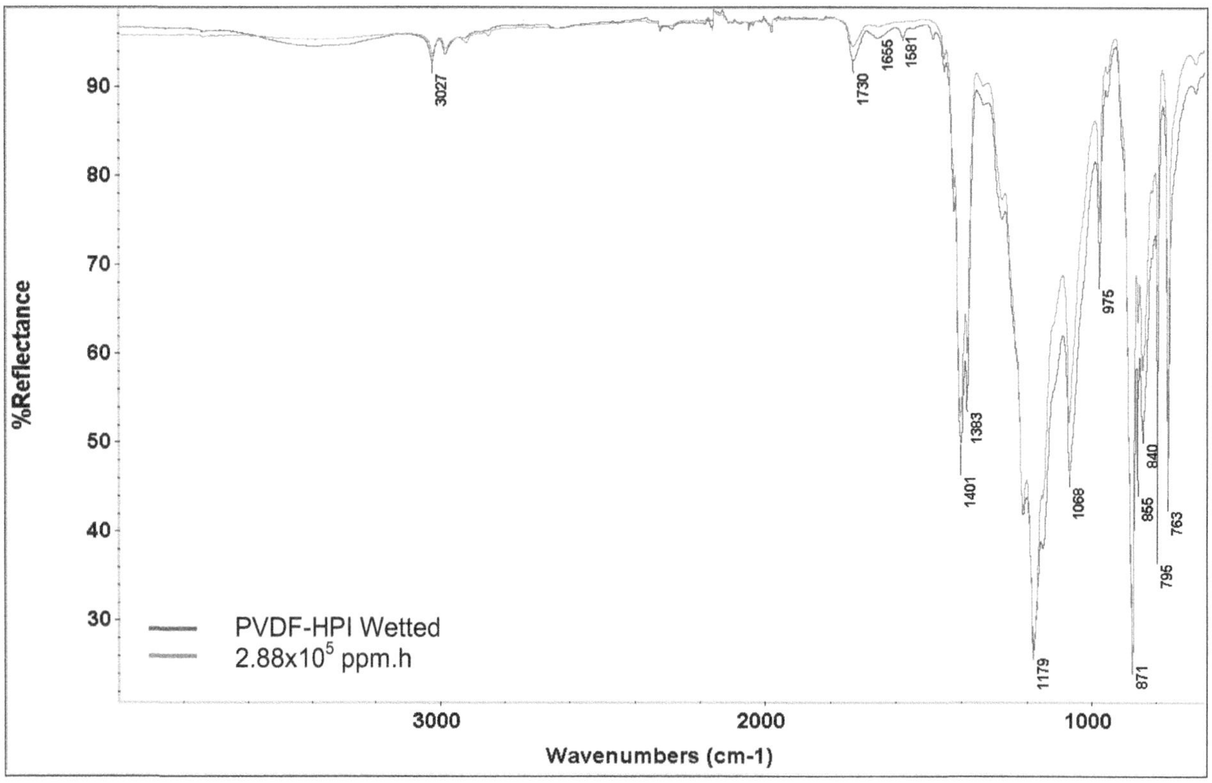

Figure C: FTIR spectrum of wetted and aged PVDF-HPI membranes in 0.04% HCl.

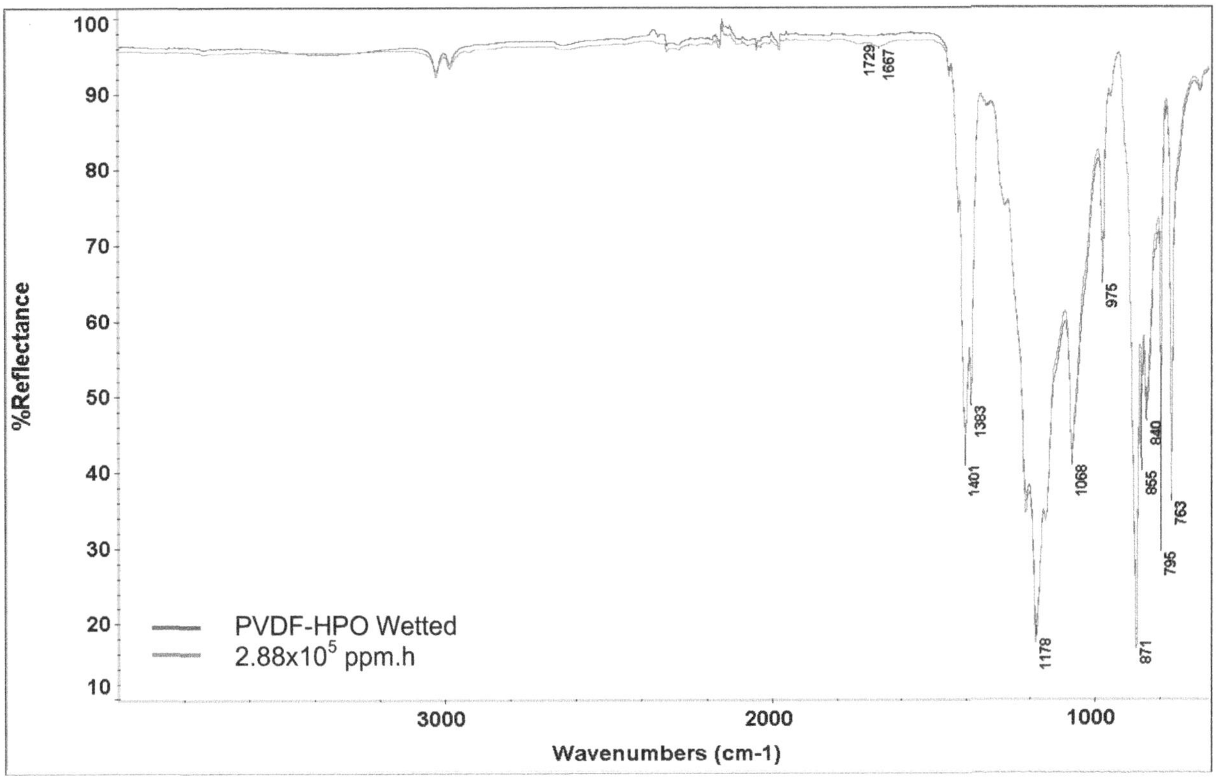

Figure C: FTIR spectrum of wetted and aged PVDF-HPO membranes in 0.04% HCl.

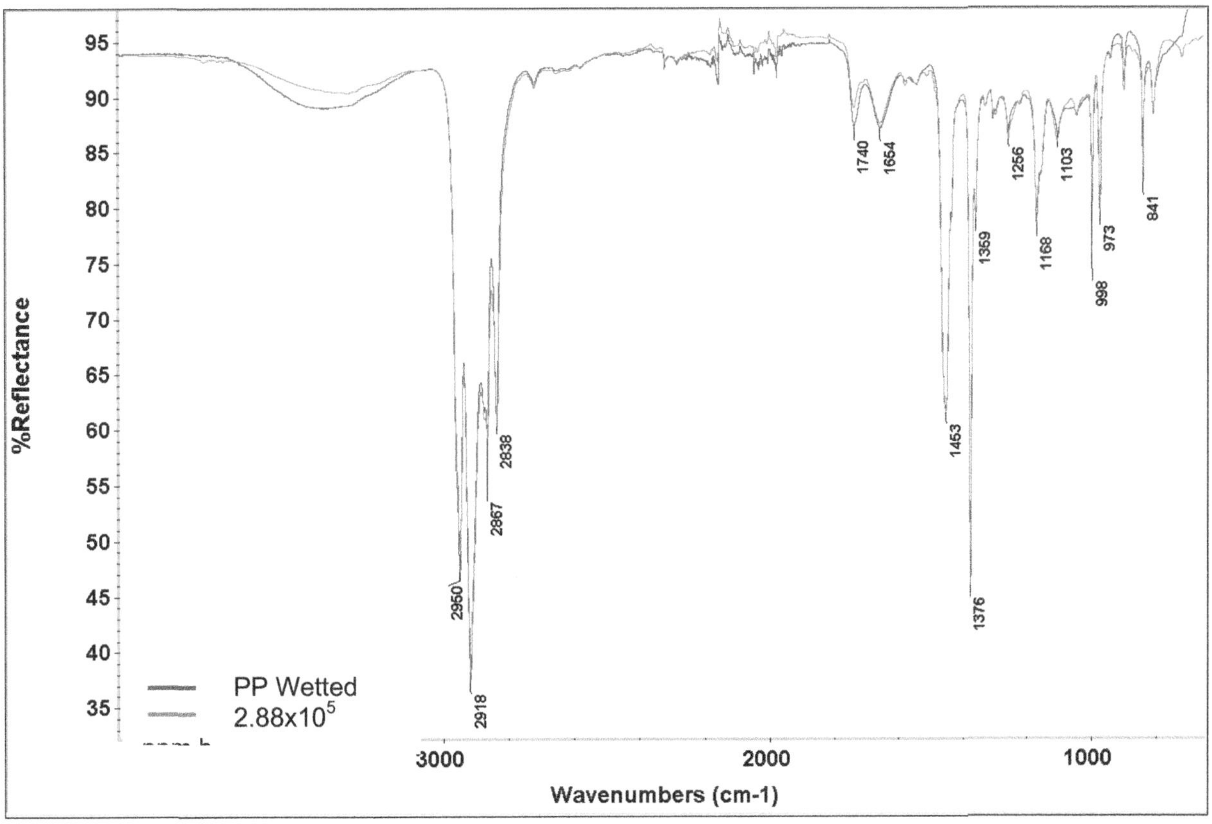

Figure C: FTIR spectrum of wetted and aged PP membranes in 0.04% HCl.

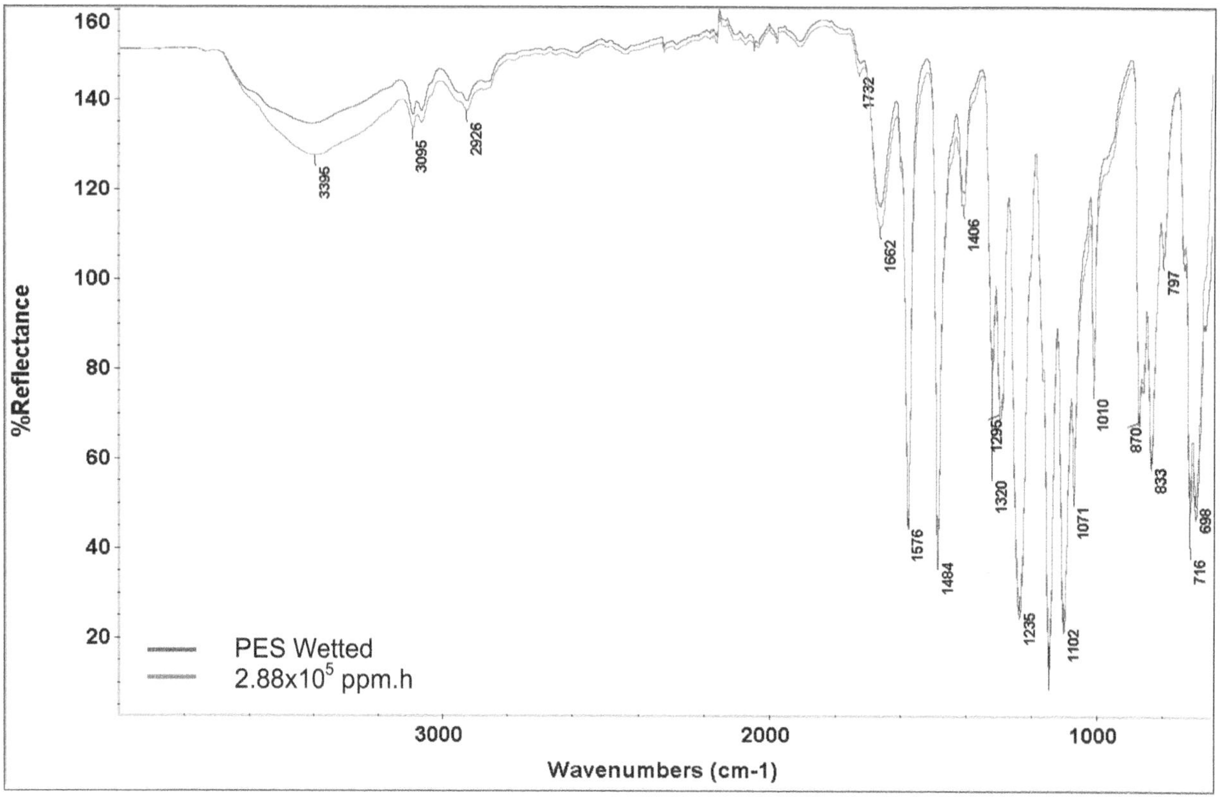

Figure C: FTIR spectrum of wetted and aged PES membranes in 0.04% HCl.

Appendix D: FTIR analysis of NaOH-aged membrane

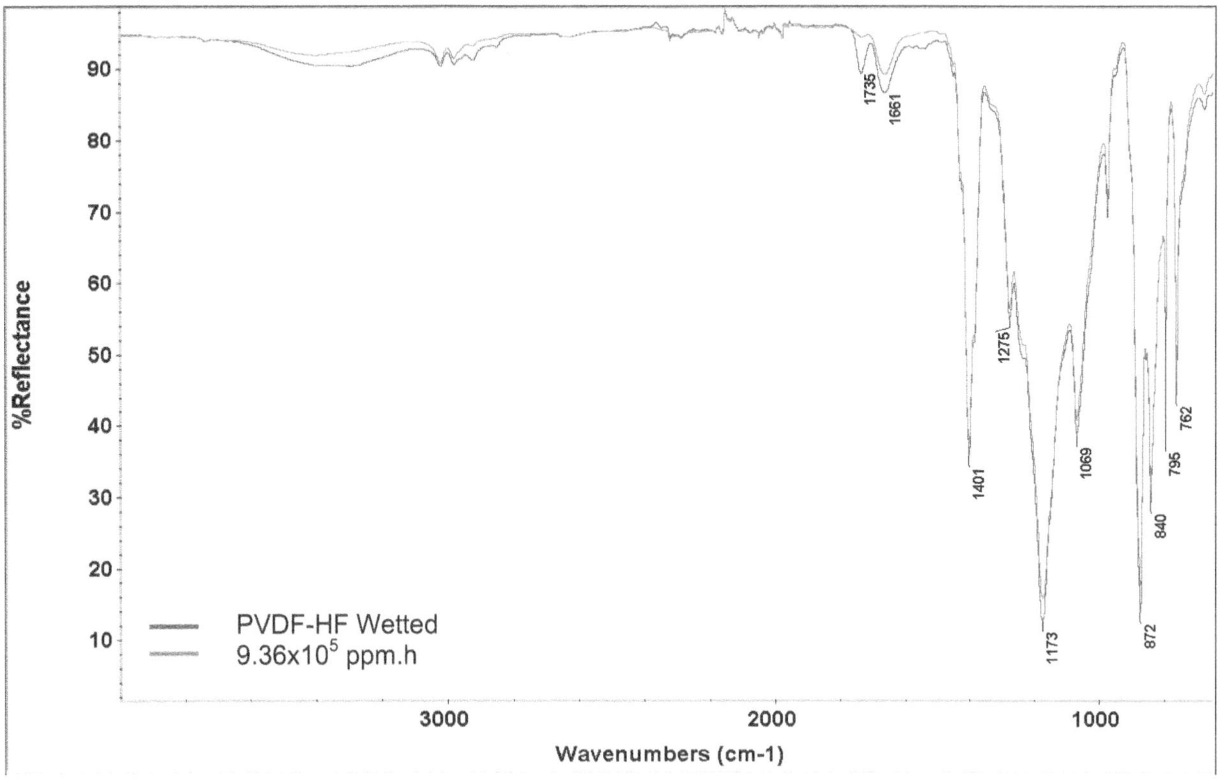

Figure D: FTIR spectrum of wetted and aged PVDF-HF membranes in 0.13% NaOH.

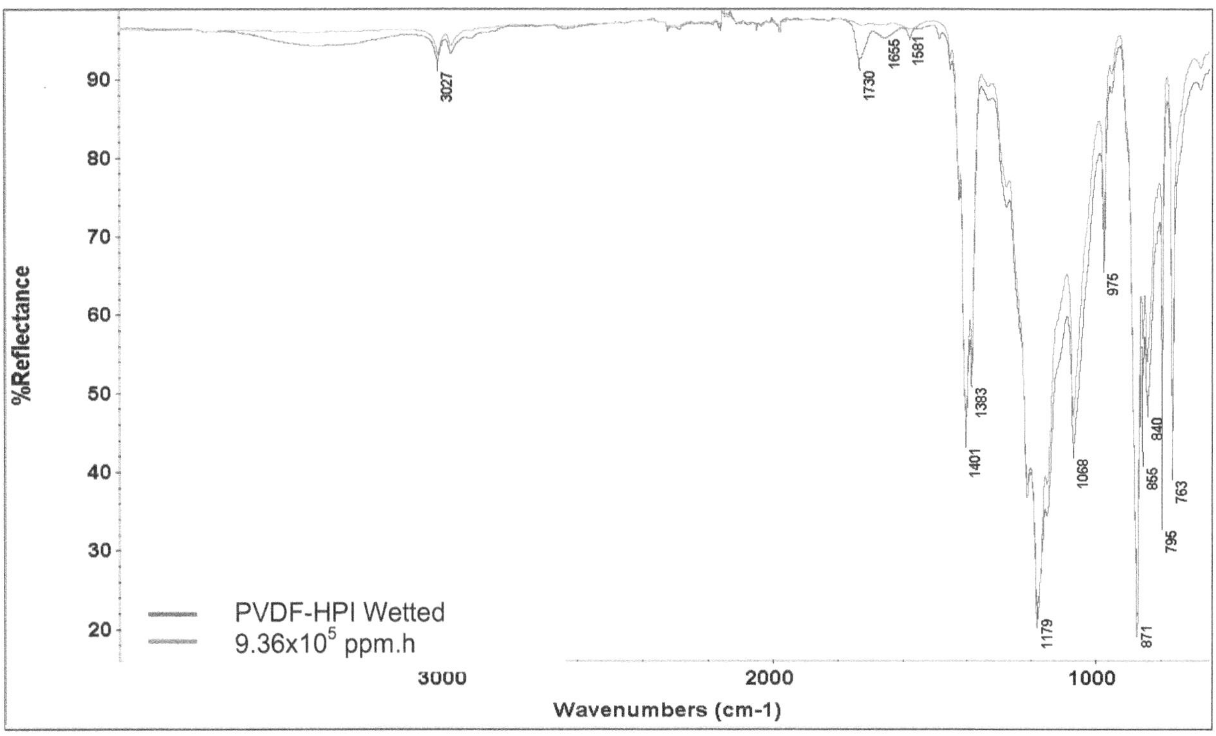

Figure D: FTIR spectrum of wetted and aged PVDF-HPI membranes in 0.13% NaOH.

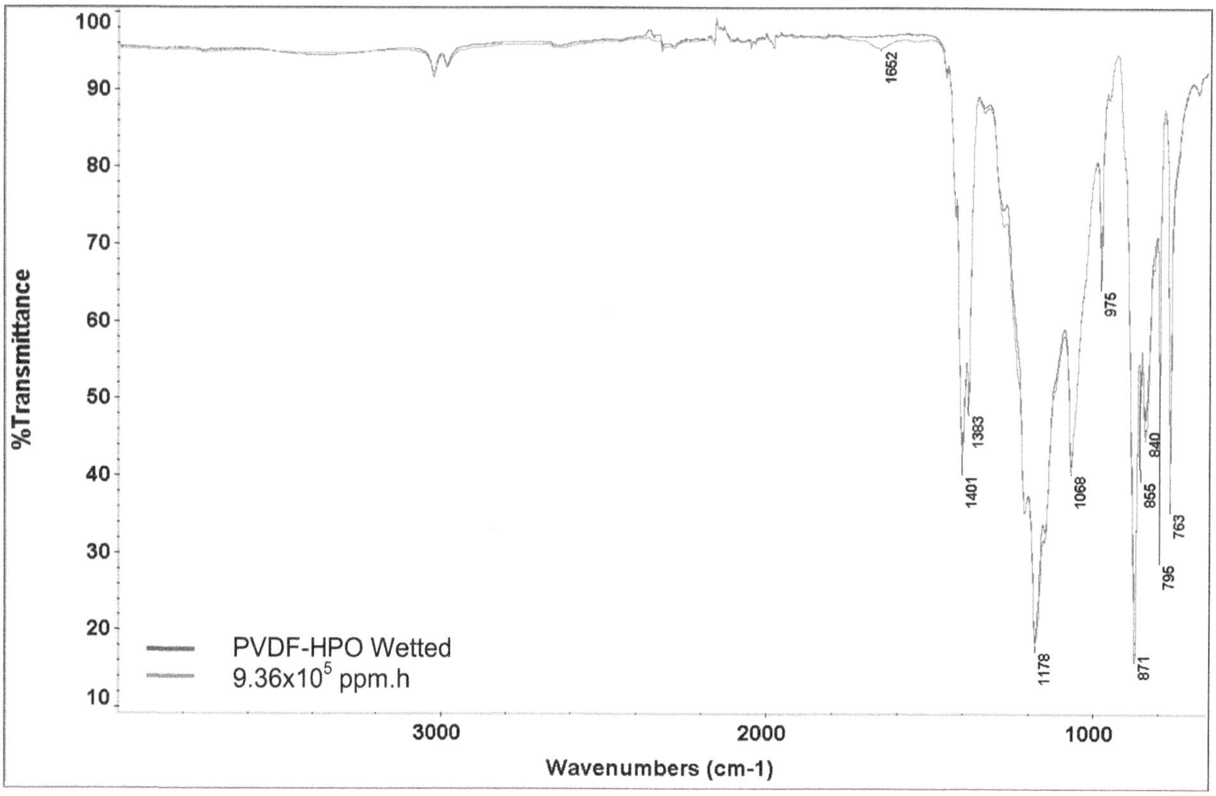

Figure D: FTIR spectrum of wetted and aged PVDF-HPO membranes in 0.13% NaOH.

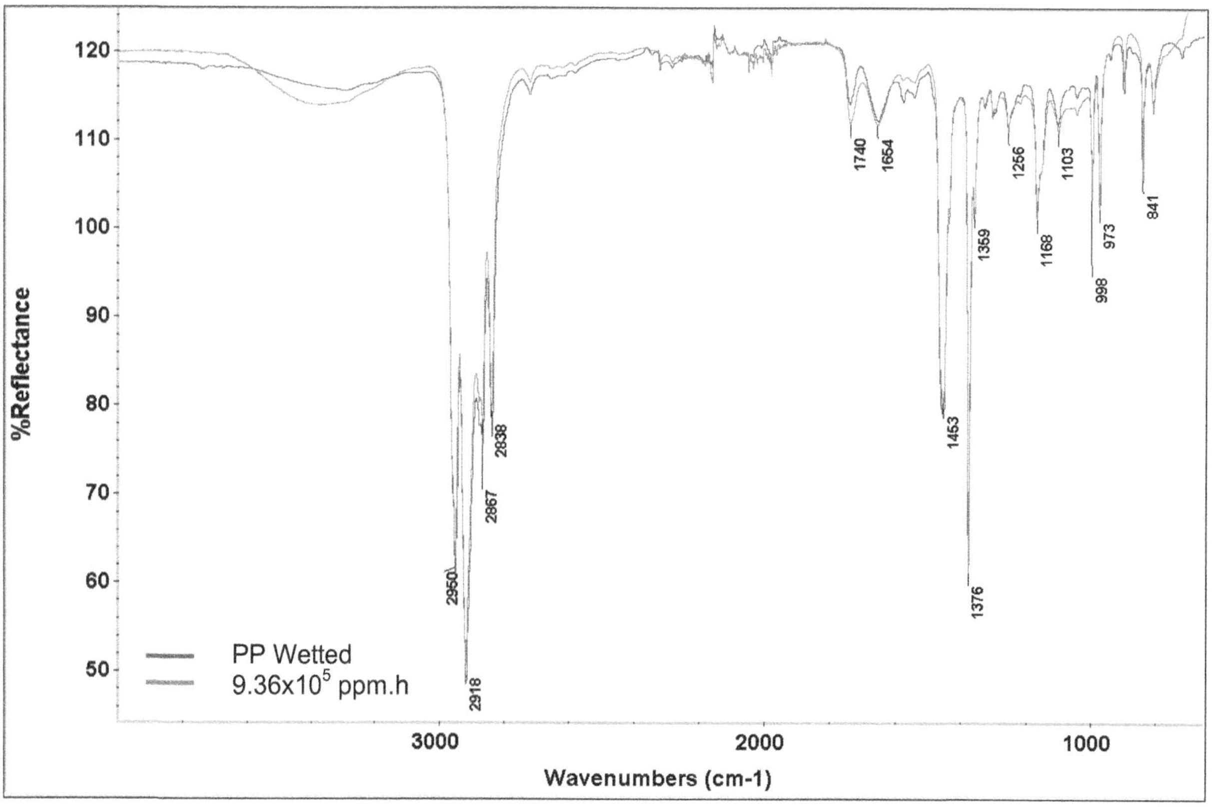

Figure D: FTIR spectrum of wetted and aged PP membranes in 0.13% NaOH.

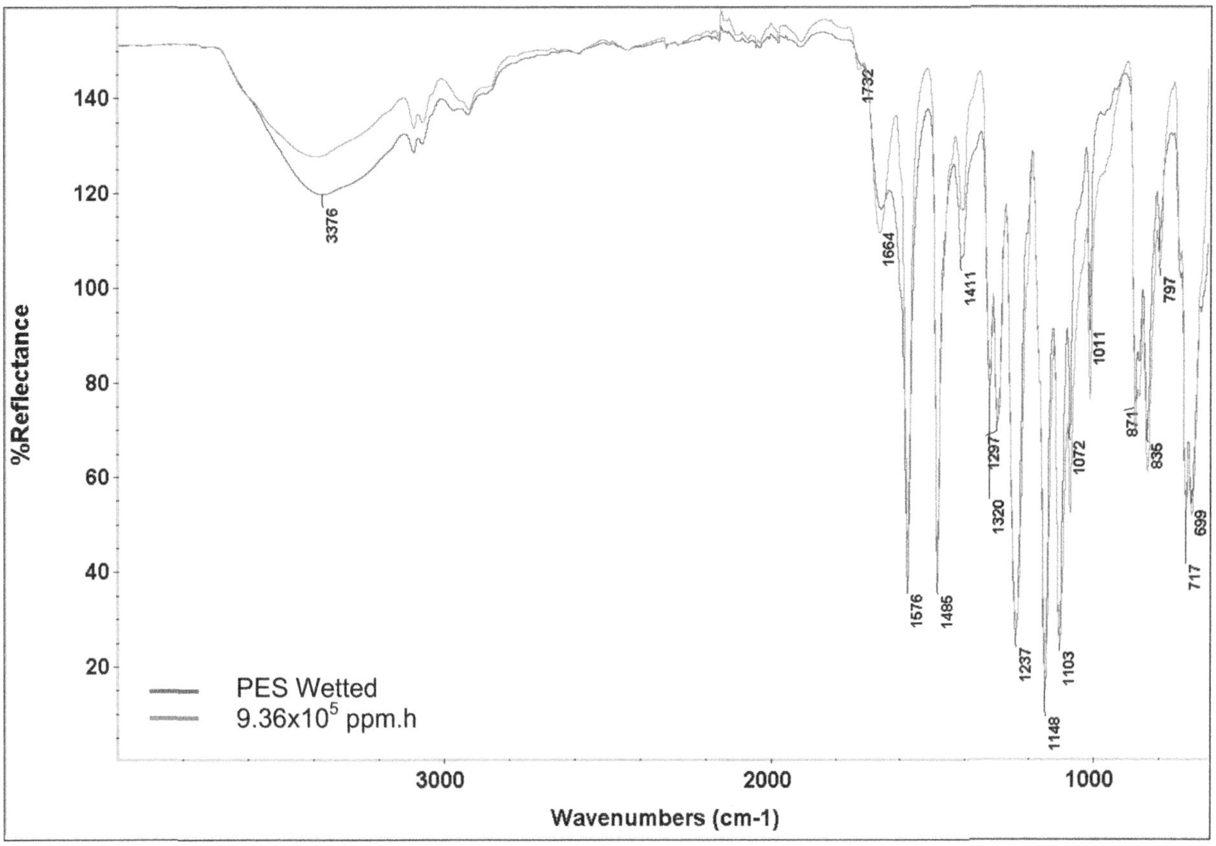

Figure D: FTIR spectrum of wetted and aged PES membranes in 0.13% NaOH.

Appendix 3

Membrane Ageing During Consecutive Fouling/Cleaning Cyclical Tests
Report No 4

Pierre LeClech

UNESCO Centre for Membrane Science and Technology, University of New South Wales

EXECUTIVE SUMMARY

Chemical cleaning is one of the most common techniques used in membrane processes to recover the performances of the system lost during filtration. However, the frequent use of chemical agents can also lead to the gradual degradation of the physical and chemical properties of the membrane, and potentially to the loss of integrity of the separation process. In this study, long-term consecutive fouling/cleaning cycling experiments were carried out to assess the nascent changes in membrane performances. The methodology developed in this research is expected to provide a more realistic assessment of membrane ageing, compared to the short-term accelerated ageing tests commonly reported in the literature. Hollow fibre polyvinylidene fluoride (PVDF) and polypropylene (PP) membranes were fouled with a mixture of humic acid, alginate and bovine serum albumin (BSA). A chemical agent commonly used in the water and wastewater industry, sodium hypochlorite (NaOCl), was selected for membrane cleaning.

Preliminary optimisation tests indicated that the use of 2% NaOCl cleaning over 90 h of filtration/cleaning cycles results in severe and measureable changes in the membrane performances. Benchmark testings were also carried out to (1) identify the most relevant membrane property capable to indicate nascent changes in structural changes and (2) assess the relative impact of fouling and cleaning on the filtration behaviour. Following the ageing process, membrane samples were characterised using contact angle, thermogravimetric analysis (TGA), Fourier transform infrared spectroscopy (FTIR), tensile and bubble point tests.

The benchmark tests revealed that longer contact time between the membrane and the cleaning agent resulted in a significantly lower membrane resistance, due to potential increase in membrane pore size. Moreover, NaOCl cleaning conducted after filtration was observed to slightly improve the overall hydraulic performances of the system. The exposure of PVDF membrane to NaOCl during cleaning cycles reduced the membrane TOC rejection performances, which was due to the potential increase in the membrane pore size or surface hydrophilicity, resulting from NaOCl impact on the membrane.

Under the operating conditions applied in this study, the impact of NaOCl on the structure of the PVDF membrane was detected by the majority of the analytical tools. The apparent pore size enhancement of the PVDF membrane was detected from both the increased permeability and hydrophilicity of the aged membrane. The removal of most of the surface modifier compounds, added to the modification of the core material of the membrane was also responsible for this increase of wettability. It was therefore proposed that the additives could play a significant role in the initial wetting of the membrane, while their presence during consecutive filtration was less crucial to sustainable operation.

Further characterisation of membrane samples indicated that NaOCl exposure resulted into the potential damage of the PVDF membrane through cross linking reaction. However, the impact of NaOCl on PVDF tensile properties was less pronounced for the fouled membrane, due to the apparent protection offered by the foulant: Oxidation of fouling materials deposited on the membrane surface was expected to occur before the degradation of the membrane material itself.

When compared to those obtained with the PVDF material, decline in PP membrane performances were significantly more pronounced. This trend was expected, as PVDF has been widely reported to be more resistant to chemical attack. After filtration/cleaning cycles on PP membrane, indications of pore enhancement were clearly observed, and ageing mechanism including chain scission resulting from the oxidisation of the polymer due to NaOCl attack was proposed. Formation of new O-H band, along with shifting and degradation of the carbonyl group (as observed during FTIR analysis of the aged membrane) confirmed this hypothesis. Moreover, the tensile properties of the PP membrane were also severely reduced after ageing.

CONTENTS

FIGURES

TABLES

ABBREVIATIONS

AFM	- atomic force microscopy
ATR-FTIR	- attenuated total reflection - Fourier Transform infrared spectroscopy
CA	- cellulose acetate
CIL	- cleaning in line
CIP	- cleaning in place
CLSM	- confocal laser scanning microscopy
EDTA	- ethylene diamine tetra acetic acid
FS	- flat sheet
GPC	- gel permeation chromatography
HF	- hollow fiber
IC	- inorganic carbon
LMH	- Lm-2h-1
MBR	- membrane bioreactors
MF	- microfiltration
NaOCl	- sodium hypochlorite
PAN	- polyacrylontrile
PE	- polyethylene
PEG	- poly ethylene glycol
PES	- polyethersulfone
PP	- polypropylene
PS	- polysulfone
PVDF	- polyvinylidene fluoride
PVP	- polyvinylpyrrolidone
Rm	- membrane hydraulic resistance
SDS	- sodium dodecyl sulphate
SEM	- scanning electron microscopy
TC	- total carbon
TGA	- thermogravimetric analysis
TMP	- trans membrane pressure
TOC	- total organic carbon
UF	- ultrafiltration
WT	- water treatment
WWT	- wastewater treatment
XPS	- X-Ray photoelectron spectroscopy

1. INTRODUCTION

Chemicals used to clean membranes in water and wastewater treatment plants do not only react with the foulants; strong evidence has indicated their participation to the degradation of the membrane material. Alterations in the membrane core polymer due to exposure to cleaning agents can result in compromised membrane rejection performance, and to membrane failure. Although polymeric degradation has been widely studied in the literature, little is known on the nascent and cumulative changes in the chemical and structural composition of porous membranes due to chemical exposure. Moreover, previous studies have been mainly based on accelerated membrane ageing techniques, which is not a realistic replicate of the cleaning procedure applied in the industry. Only few studies have examined cyclical cleaning effects on membrane performance [1-2]. In this study, the physical and chemical characteristics of membrane were assessed during cyclical fouling/cleaning operation to compare the accelerated ageing process in a more realistic environment. The design of the proposed experimental plan is also expected to characterise the relative impact of fouling and cleaning on the filtration behaviour.

In this study, a common oxidizing cleaning agent (sodium hypochlorite, NaOCl) was used during consecutive fouling and cleaning cyclical runs. By using a range of carefully selected analytical techniques (from previous work package), the complete assessment of the membrane nature was carried out after ageing. This study will particularly focus on the application of a rigorous methodology to allow more detailed understanding of the long-term performances of membrane systems, for a range of membrane materials.

2. MATERIALS AND METHODS

2.1 Membranes

The study was based on micro-porous hollow fibre polyvinylidene fluoride (PVDF) and polypropylene (PP) membranes. The hydrophilic PVDF hollow fibre membrane featured average pore size of 0.04 μm with outer and inner diameters of 0.8 and 0.5 mm, respectively. Hollow fibre PP membranes featured pore diameter of 0.2 μm, outer and inner diameters of 0.65 and 0.39 mm respectively. Both membranes were supplied by Siemens.

2.1.1 Module preparation

Tubular cross-flow membrane modules provided the opportunity to use fresh cleaning solution, and also maintain the feed solution concentration relatively constant during the experiments. These membrane modules were composed of 6 parts:

- Two short PVC tubes (±6 cm) with 12.8 mm outside diameter (OD),
- Two union tees connecters,
- One long (9.6 mm ID, 12.8 mm OD) transparent pipe (length depends on demand),
- Hollow fibre membrane, which length should be 3-4 cm longer than the total length of module for potting purpose.

The fibres were inserted inside the transparent tube, where both sides of the fibres were glued using Araldite inside the short pipes. The tubes were then cut at the position shown below and the end of fibres at one side was blocked using glue. The module was then cut T the black line to produce adequate permeate channels (

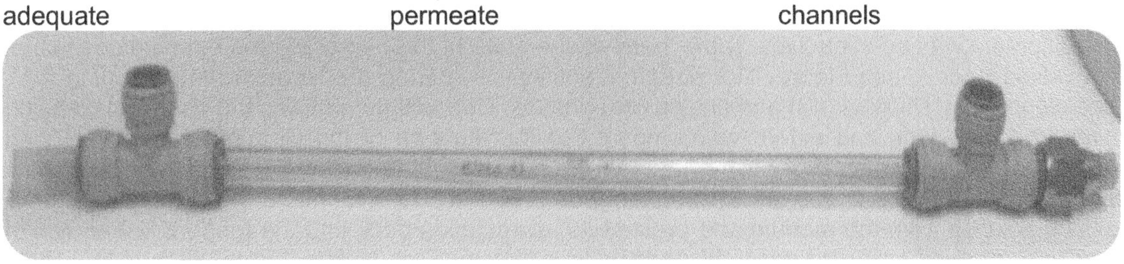

Figure 1).

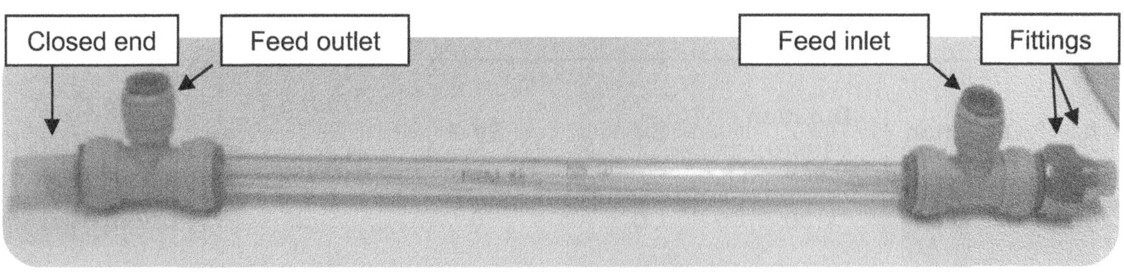

Figure 1- Picture of the hollow fibre module used in the cycling experiments.

2.2 Chemicals

Sodium hypochlorite (NaOCl) from Ajax Finechem with nominal concentration of 12.5% w/v was used as the chemical cleaning agent in the cyclical runs. Humic acid sodium salt, alginic acid sodium salt and bovine albumin serum (BSA) were purchased from Sigma Aldrich. Humic acid, alginate and BSA were used to prepare the feed solution at concentrations of 20, 10 and 10 mg/L respectively.

2.3 Filtration set up and automatic operational phases

In this series of experiments, the membrane fouling/cleaning operation was conducted using the filtration set up shown in Figure 2. The setup was composed of two pumps, two 3-way solenoid valves

(SV), one pressure gauge, one pressure transducer, one membrane module and four solution vessels.

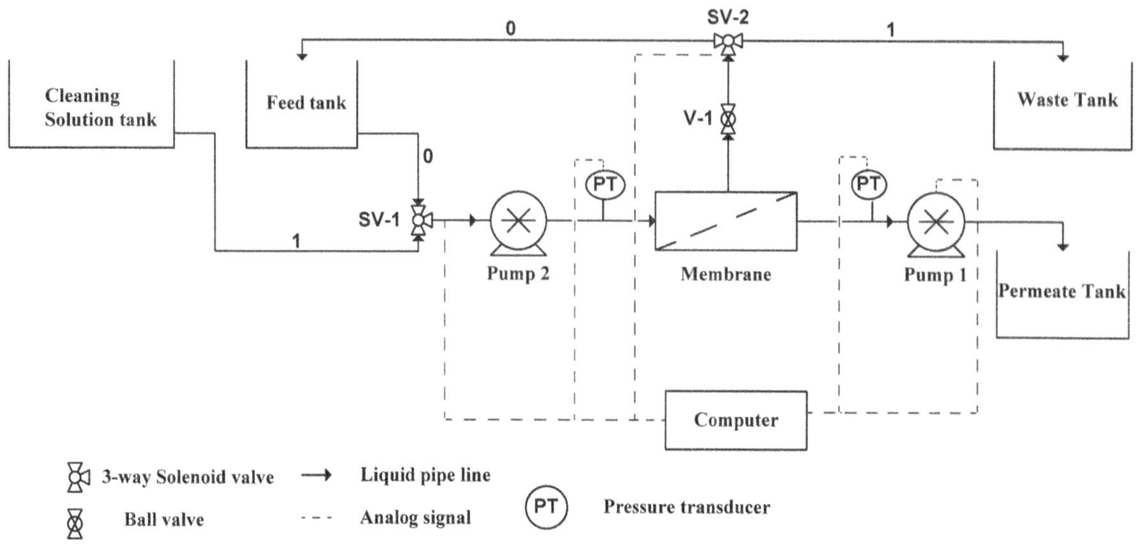

Figure 2- Schematic illustration of the cyclical fouling/cleaning setup.

All the equipment was pc-controlled. Table 1 shows the state of the pump and the opening position of the two valves (0 or 1, see Figure 1 for stream identification) during the filtration (#1), cleaning (#2), assisted cleaning by filtration (#3) and rinsing (#4) phases. Due to time restrain, the chemical cleaning time was artificially long and extensive, compared to the duration of the filtration. A relatively long rinsing period was also necessary to remove all trace of NaOCl before the filtration was resumed. During rinsing time, Pump 2 was set to a higher speed (60 L m^{-2} hr^{-1}) to insure the complete removal of cleaning solution from the module and prevent its recirculation back into the feed tank. The flux of the feed stream was adjusted manually and was maintained at constant rate during the whole operation. Feed solution was used to rinse the cleaning solution from the module after cleaning phase. Permeate flux rate was fixed at 30 or 50 L m^{-2} hr^{-1}.

Table 1- Operational phases and position of the solenoid valves during each fouling/cleaning cycle.

#	Operation	Duration (min)	SV-1	SV-2	Pump 1 (Permeate)	Pump 2 (Feed)
1	Filtration	30	0	0	on	on
2	Cleaning	17	1	1	off	on
3	Assisted cleaning	2	1	1	off	on
4	Rinsing	10	0	1	off	on

Blank experiments (i.e. using MilliQ water) were based on the same protocol to assess the relative impact of the different mechanisms which could influence membrane properties. This systematic experimental approach can expand the current understanding of the factors involved in membrane ageing. The different experiments carried out in this study are shown in Table 2:

Table 2- List of experiments conducted in this study and their identifications.

Identification	Foulant	Cleaning solution
MQ/MQ	Milli-Q	Milli-Q
MQ/NaOCl	Milli-Q	NaOCl (17 min/cycle)
MQ/2×NaOCl	Milli-Q	NaOCl (32 min/cycle)
Foulant/MQ	Feed	Milli-Q
Foulant/NaOCl	Feed	NaOCl (17 min/cycle)

2.4 Membrane characterisation

Following the ageing process, different membrane characterisation techniques were adopted to investigate the suitability of the each method in assessing membrane degradation. In the previous study (work package #2), the suitability and reliability of the different membrane characterisation techniques in detecting membrane degradation were assessed and compared. The comparison was performed by developing a ranking system, with each criterion being given a weighting factor depending on its importance.

Table 3 shows the calculated reliability of the selected membrane characterisation techniques. These tests were carried out on five new samples of 0.04 µm HF/PVDF membranes. The average value of the samples was calculated for each test. The sample standard deviations, s, of each characterisation method were then computed to find the variability in results from the mean value.

Table 3- Summary of the reliability assessment of analytical techniques used on new PVDF membrane.

Characterisation techniques		Standard deviation (%)
Clean water test		6
Bubble point test		20
Contact angle		19
Tensile test	Breaking force	3
	Ultimate elongation	10
TGA		1

2.4.1 Membrane resistance and TMP behaviour

Clean water test was performed after wetting and rinsing the membrane to measure its intrinsic resistance (R_m). The clean water test for the hollow fibre membranes in the tubular module was performed using the flux stepping method. Transmembrane pressure (TMP) was calculated as the difference in pressure between the feed and the permeate side of the membrane measured by pressure transducers. Constant flux filtration was carried out at 20 $Lm^{-2}h^{-1}$, and increased to 50 and 100 $Lm^{-2}h^{-1}$ after 30-60 s intervals, while the pressure was recorded for each flux step.

Figure 3 illustrates the TMP behaviour of hollow fibre membrane during a typical filtration and cleaning run. During the fouling/cleaning experiments, the TMP behaviour during each filtration cycle was expressed as fouling rate (dTMP/dt) and was calculated using Equation 2:

$$\frac{dTMP}{dt} = \frac{TMP_b - TMP_a}{t_b - t_a} \quad \text{Equation 2}$$

Where TMP_a and TMP_b are the initial (after 10 min of filtration) and final TMP (kPa) of the filtration cycle, respectively. The t_a and t_b as their corresponding times (h). The overall TMP increase rate during the entire filtration time of these experiments ($dTMP/dt_{overall}$) was calculated using Equation 3.

$$\frac{dTMP}{dt}_{overall} = \frac{Average(TMP_d, TMP_e, TMP_f) - Average(TMP_a, TMP_b, TMP_c)}{t_f - t_a} \quad \text{Equation 3}$$

Where $TMP_{d\ to\ f}$ and $TMP_{a\ to\ c}$ are representing the final TMP of the last and first 3 three filtration cycles.

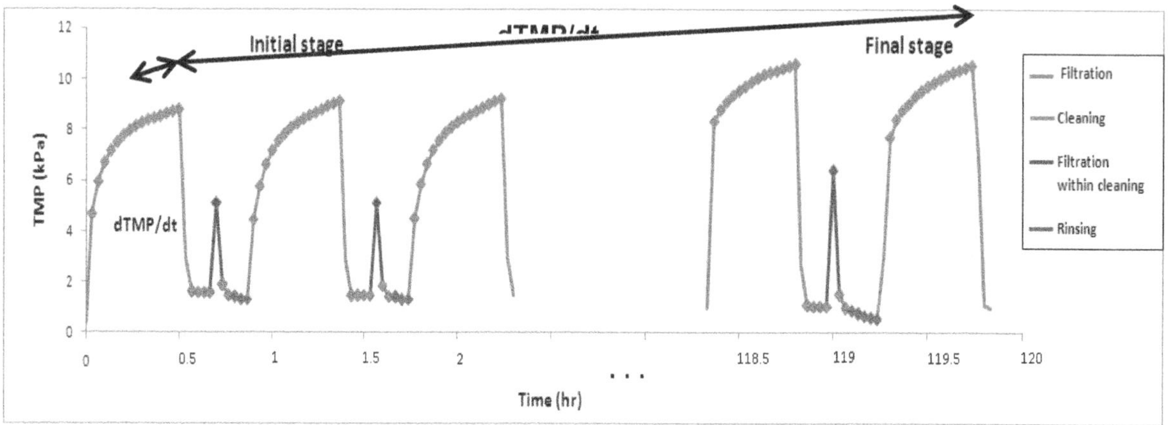

Figure 3- TMP behaviour and fouling rates (dTMP/dt) calculations during initial and final stages during fouling/cleaning cyclical experiment.

2.4.2 Tensile test

Measurements of the mechanical properties of the membranes were performed with tensile tests. The material was clamped at both ends and stretched at constant speed. The force required to elongate the material with time was recorded until the material breaks. The tests were performed on a TA.XT2i/5 texture analyser (Texture Technologies), which has a maximum pulling force of 50 N. Hollow fibre samples were cut into 40 mm length. The test speed and the initial distance were 0.5 mm/s and 35 mm respectively. The software used to collect information from the Texture Analyser is Stable Micro Systems® Texture Expert. It recorded the force needed to elongate the material and the distance between grips per sampling time and plotted the graph of force (N) against time (s).

2.4.3 Thermogravimetric analysis (TGA)

Thermogravimetric analysis (TGA) is a thermal analysis technique which measures the weight change in a material as a function of temperature and time in a controlled atmosphere. The technique is useful when investigating the behaviour and thermal stability of materials, as well as performing compositional analysis of materials. Qualitative and quantitative analysis could be carried out with minimal solid sample size, with measurement range between 0.1 and 200mg. TGA tests in this study were performed on the Thermal Advantage® device manufactured by TA Instruments. The samples were dried prior to testing to prevent the presence of moisture in the samples. Analysis was carried out in inert atmosphere (N_2 gas), with temperature increase in the range of 20-1000°C. Test method chosen was the Ramp method, which increases the temperature 20°C/min.

2.4.4 Fourier transform infrared spectroscopy (FTIR)

FTIR allows the detection of changes in the functional groups of the membrane by comparing membranes before and after exposure to cleaning agents. FTIR detects bonds between atoms in a molecule and can thus provide information about functional groups and chemical structures of the membrane. Each peak illustrates a specific type of molecular vibration: symmetric, asymmetric, twisting, wagging, bending or rocking. The FTIR spectra were analysed with the high resolution

Hummel polymer and additives spectrum library available with the FTIR program. Bruker IFS66/S FT/IR Spectrometer was applied to obtain the spectrums for new and aged membranes.

2.4.5 Bubble point test

To determine the bubble point pressure of the hollow fibre membranes, the potted modules were completely filled with Milli-Q water. Then, nitrogen gas was connected to the feed inlet side of the module. The pressure was read from a pressure indicator attached between the gas cylinder and the membrane module. During the test, the pressure was gradually increased until the first bubble was observed through the membrane surface.

2.4.6 Contact angle

The contact angle measurement of the new and aged membranes was performed with the KSV 200 CAM goniometer, with Milli-Q water was used as the reference solvent. The meniscus measurement technique was adopted for the tested hollow fibre membranes.

The suggested procedure of measuring contact angle of cylindrical shapes such as rods and fibres with the KSV CAM 200 Goniometer is the meniscus measurement method. The recorded contact angle images for hollow fibre membranes needed to be manually adjusted. From the curve fitting window, the liquid baseline, meniscus/air boundary and solid/air boundary were fixed. The contact angle values were then calculated by the software using the average of the left and right meniscus values.

3. RESULTS AND DISCUSSION

3.1 Physicochemical properties of membranes during wetting and rinsing

In this initial series of experiments, the impact of membrane preparation (i.e. wetting and rinsing) on its properties was assessed. As mentioned earlier, hollow fibre membranes are required to be wetted in ethanol solution to gain permeability. In this study, the membranes were immersed in 90%v/v ethanol for a total duration of 2 hr. The membrane was then rinsed and Milli-Q water was filtered through the membrane at 50 L/m^2hr, while permeate samples were analysed in terms of total organic carbon (TOC). High TOC concentrations were detected following membrane wetting process due to the presence of ethanol residual on the membrane. It was also expected that other preservative and additive organic compounds were present in the rinsing solution. Therefore, the modules were rinsed extensively to eliminate the remaining ethanol form the membrane. It was observed that 2 hr of filtration were necessary for the TOC content of the permeate to reduce below the acceptable concentration of 2 mg/L.

In this study, a blank experiment (identified as 'MQ/MQ', where Milli-Q was used as feed and cleaning solutions) was carried out to obtain a baseline for the PVDF membrane long-term filtration behaviour, without the impact of fouling and/or cleaning process. This step also allowed to confirm that the hydraulic performances of the wetted and rinsed membranes were stable, as indicated by the constant TMP obtained during the week of testing.

To further investigate the impact of membrane preparation, the TGA and FTIR characterisation results from the MQ/MQ experiment and the virgin and the wetted membrane samples were compared. Figure 4 shows the rate of mass loss as a function of temperature for virgin, wetted and MQ/MQ membranes. For virgin membrane, three peaks were observed during the thermal degradation process; whereas, only two types of compounds were detected for the wetted and MQ/MQ membranes.

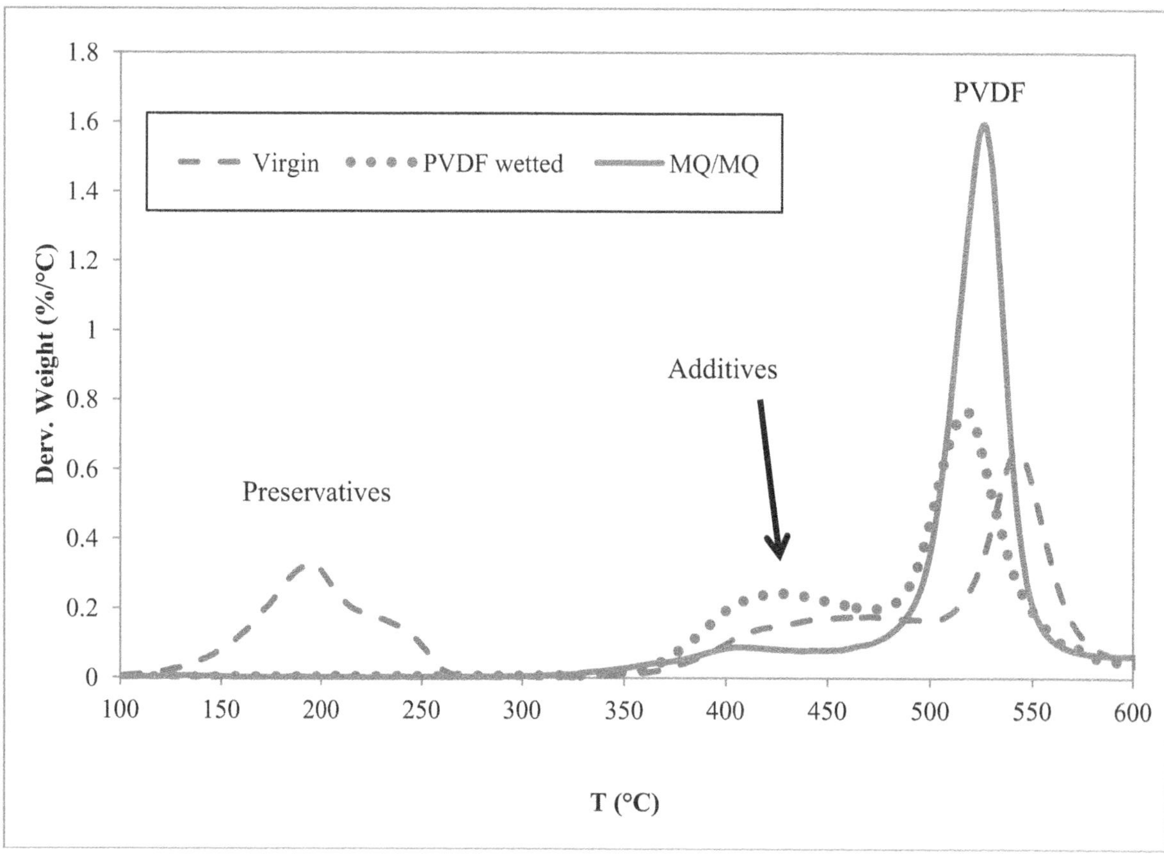

Figure 4- Thermal properties of virgin, wetted and MQ/MQ PVDF membranes by TGA.

Commercially available virgin membrane normally contains preservatives such as sodium (meta- or not) bisulfite, which are added by the manufacturers to increase the shelf-life of the product by preventing biological growth on the membrane [3]. Given their thermal properties, the preservatives are expected to be degraded between 150 and 250°C; this confirms the presence of these compounds in the virgin membrane and their complete removal after the wetting process. This stage was not observed during the thermal degradation of the other membranes, since membrane preservatives were eliminated during the rinsing step with Milli-Q water. In all samples, the peak observed between 360 and 480 °C (24% mass loss) corresponded to thermal degradation of polyvinylpyrrolidone (PVP), which has been reported as a common membrane additive. Additives in casting solution are widely used for pore-forming and/or surface modification, in order to create a more hydrophilic membrane [4-5]. The last degradation peak clearly indicated the presence of PVDF, which his typical peak appearing between 480-580 °C [5].

The typical PVDF and PVP peak values and their corresponding functional groups are summarised in Table 4. In order to confirm the presence of PVP in the PVDF membrane, FTIR spectrum of both pure materials were analysed and compared. As mentioned previously, the presence of PVP was confirmed in the membrane polymer and, the 1662 cm^{-1} peak was found to be the best indicator to identify this compound in the PVDF membrane samples. the missing carbonyl peak at 1729was on the Durapore flat sheet PVDF membrane was related cm^{-1} to the polyamine or polyamide-polyamine epichlorohydrin resin surface modifier, in a previous study [6]. Epichlorohydrin is an epoxy resin applied as a surface modifier to increase the membrane hydrophilicity and charge [7].

Table 4- PVDF and PVP structure, FTIR peak values and their corresponding functional groups [6, 8].

Core membrane materials	Structure	Peak value (cm^{-1})	Functional groups
PVDF	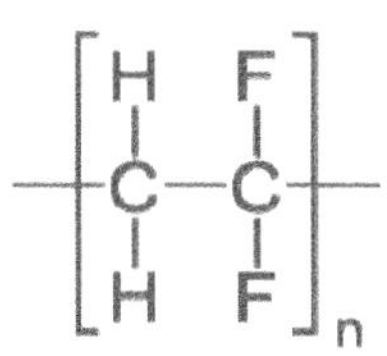	763	PVDF skeletal bonding
		975	PVDF fingerprint
		1069	PVDF fingerprint
		1732	Carbonyl
PVP	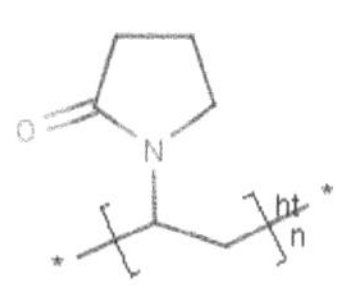	1662	Carbonyl

Figure 5 shows the FTIR analysis of virgin, wetted and MQ/MQ PVDF. FTIR analysis of these membranes spectrums exhibited two absorbance peaks between 1500 and 2000 cm^{-1}, which are not present in pure PVDF spectra. Wavenumber 1664 corresponds to presence of PVP [9], which was also detected by TGA. The other peak at 1734 was most likely another membrane additive. The carbonyl peak at 1729 cm^{-1} can be related to the polyamine or polyamide-polyamine epichlorohydrin resin which is also a surface modifier [6]. Virgin membrane spectrum demonstrated additional peaks between 900 and 1100 cm^{-1}, confirming the presence of preservatives in the sample [10].

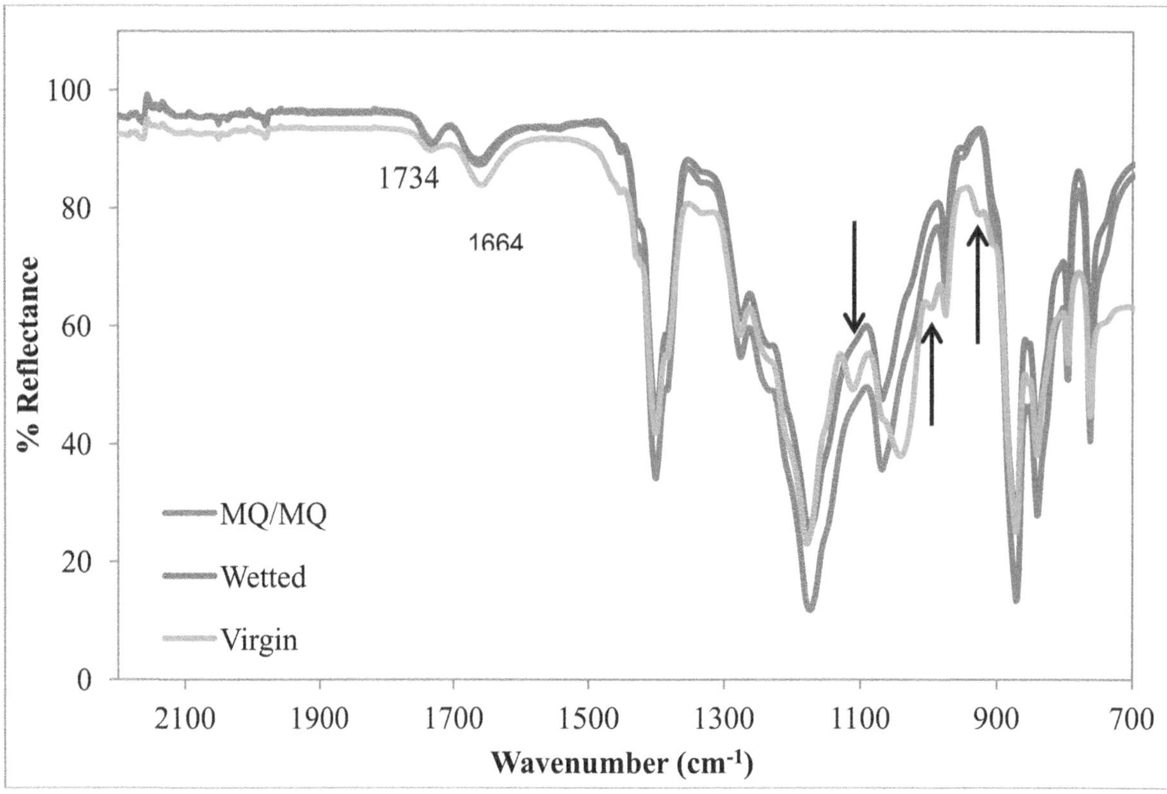

Figure 5- FTIR spectra of virgin, wetted and MQ/MQ hollow fibre PVDF membranes.

These results implicated that the preservative solution was removed from the sample after wetting with ethanol and then rinsing with Milli-Q. The peaks which represented the additives in the membrane blend were clearly confirmed from the FTIR analysis of these samples. It was also observed that relatively long-term filtration of Milli-Q water (MQ/MQ) did not influence the carbonyl peaks (1664 and 1734 cm^{-1}) of the wetted sample. This indicates that although PVP is water soluble, it is resistant to membrane rinsing. Similarity of spectrum peaks for virgin, wetted and MQ/MQ samples indicated that membrane preparation did not have a significant impact on the overall membrane nature and the physically attached additives such as PVP and epichlorohydrin.

3.2 Initial testing for ageing optimisation

In this study, initial experiments using two different NaOCl concentrations (0.04 and 2%) and permeate fluxes were performed to select the optimum experimental protocol. Factors such as fouling rate and TOC rejection were measured during the initial and final stages of the experiment.

Membrane Ageing During Consecutive Fouling/Cleaning Cyclical Tests

Table 5 shows the preliminary cyclical fouling/cleaning experiments results for hollow fibre PVDF membrane. The reported dTMP/dt values in Table 5 are the average of three consecutive initial and final filtration cycles of the experiment. The standard deviation in terms of the TOC rejection tests were less than 2%, measured through TOC analysis of triplicate humic acid standard (10 mg/L) samples. The cleaning intensity of NaOCl was calculated and reported in ppm.h.

Table 5- Preliminary cyclical fouling/cleaning tests and the obtained initial and final fouling rates and TOC rejection values using hollow fibre PVDF membrane.

Flux (L/m^2hr)	**NaOCl (ppm.hr)**	**dTMP/dt $_{initial}$ (kPa/hr)**	**Rejection $_{initial}$ (%)**	**dTMP/dt $_{final}$ (kPa/hr)**	**Rejection $_{final}$ (%)**
30	102,000	1.0	87	5.4	80
30	510,000	1.0	87	2.6	65
50	102,000	1.7	87	44.2	98
50	510,000	1.6	87	30.5	69

In the initial stages of the experiment, higher fouling rate was observed at 50 L m^{-2} hr^{-1} permeate flux. The initial TOC rejection of PVDF membrane was 87% for all four experiments. However, the membrane performance altered significantly at the final stages of these tests. The TOC rejection of PVDF membrane was observed to be lower than its initial value in all experiments.

At 30 L/m^2hr permeate flux using the high concentration of NaOCl (500,000 ppm) in cleaning cycles resulted in lower overall fouling rate. The TOC rejection of the PVDF membrane was reduced more significantly, where 500,000 ppm NaOCl was applied. Similar trends were observed, when the permeate flux of 50 L/m2hr was applied however, the changes in membrane performances were more pronounced.

Since the duration of these tests was only 90 hr, NaOCl concentration of 510,000 ppm.hr (NaOCl 2% and 17 min/hr cleaning time) was selected which corresponds to 25.5 hr of accelerated ageing. This concentration and total cleaning duration allows a fair comparison between the results from this study and data from the literature for static accelerated aged membranes. Moreover, the flux of 50 L/m^2hr was selected to assess the impact of cleaning agent on PVDF membrane under sever fouling conditions. Larger alterations in the membrane performances can allow the detection of the occurrence of nascent events leading to membrane damage.

3.3 PVDF membrane ageing

In this section, the results from consecutive fouling/cleaning experiments using PVDF hollow fibre membrane are discussed. These results include TMP behaviour, contact angle analysis, TGA, FTIR, tensile properties and rejection, as well as bubble point pressure.

3.3.1 TMP behaviour

The TMP temporal changes against filtration time during consecutive fouling/cleaning cycles using feed and 2% NaOCl solution, as well as the various benchmark experiments are summarised in Figure 6. By gradually using NaOCl and synthetic fed solution for fouling (i.e. mixture of BSA, humic acid and alginate) instead on Milli-Q, this series of tests was designed to assess the relative impact of fouling and cleaning to the filtration behaviour. Each data point in the Figure 6 is an average of recorded TMPs during 2 min. In all experiments, the initial TMP was comprised between 10 to 15 kPa and each filtration cycle run 30 min.

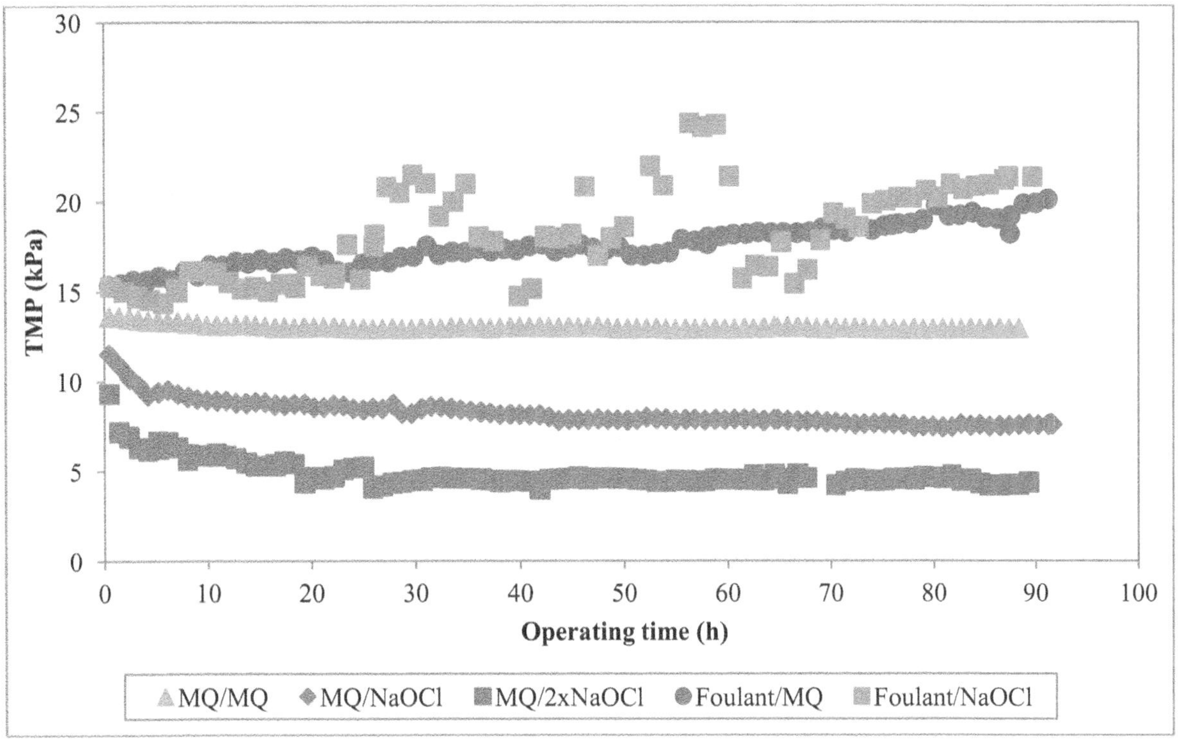

Figure 6-Temporal changes in TMP during filtration cycles (PVDF membranes).

Table summaries the TMP values and the results from the hydraulic and rejection performance testing performed prior and after each experiment. dTMP/dt values in Table 6 represents the average of the last 3 filtration cycles of the experiments. Moreover, the Rejection test was carried out at the end of the experiment. Although R_m remained relatively constant across the range of the experiments, the initial TMP behaviour (within few cycles) was strongly affected by the operating conditions. During the MQ/MQ experiment, the TMP was almost constant at 12 kPa during the overall filtration time. A slight decrease of the fouling rate in individual cycles was observed (0.21 kPa/hr), while the overall dTMP/dt of the experiment decline was negligible. This was probably due to further wetting of the PVDF hollow fibre membrane during the experiment. These results were considered as the initial benchmark indicating the behaviour of the membrane without the impact of neither the foulant nor the cleaning solution.

Table 6- Hydraulic and rejection performances for PVDF membrane during fouling/cleaning cyclical experiments.

Foulant/ Cleaner	$R_m \times 10^{-11}$ (m^{-1})	$R_{final} \times 10^{-11}$ (m^{-1})	TMP_{final} (kPa)	dTMP/dt (kPa/hr)	dTMP/dt $_{overall}$ (kPa/hr)	Rejection %
MQ/MQ	10	9.7	13	-0.21	-0.01	90
MQ/NaOCl	8.7	6.1	8	0.17	-0.04	57
MQ/2×NaOCl	6.6	3.0	4	0.22	0.01	65
Foulant/MQ	8.5	18.6	29	12.65	0.10	88
Foulant/NaOCl	7.9	8.3	24	31.23	0.06	69

During the MQ/NaOCl and MQ/2×NaOCl experiments, the TMP decreased significantly within the first three fouling/cleaning cycles. Then, the TMP continued to decrease gradually during the remaining of the filtration time from 13 to 8 kPa. This decline was more pronounced (from 11 to 4 kPa) where the cleaning cycles duration doubled (i.e. during the MQ/2xNaOCl test). Therefore, longer contact time between the membrane and the cleaning agent resulted in a significantly lower R_m, due to potential increase in membrane pore size, following alterations of PVDF polymer by NaOCl. As expected, the presence of the BSA, humic and alginate mixture in the feed solution resulted in the significant increase of the TMP_{final} (from 13 to 29 kPa) and $dTMP/dt_{overall}$ (from 0 to 0.10 kPa/h). When NaOCl was used as cleaning agent during filtrations with foulants (Foulant/NaOCl experiment), the overall hydraulic performances of the system was observed to slightly improve, with TMP_{final} of 24 kPa.

At the of each experiment, a rejection test was also preformed, using the same feed water, as described in Section 2.2. The exposure of PVDF membrane to NaOCl during cleaning cycles reduced the membrane TOC rejection performances from 90 to 57%. This decrease in TOC rejection can be also explained by the potential increase in the membrane pore size or surface hydrophilicity resulting from NaOCl impact on to the membrane. When membrane was fouled, the cleaning agent was mainly responsible for oxidising and therefore removing the fouling layer. As a result, the contact between the membrane polymer and the cleaning agent was limited and the extent of decrease in rejection was considerably lower when fouling materials were used. Therefore, NaOCl direct attack to the membrane was responsible for the significant decrease in membrane TOC rejection.

3.3.2 Contact angle

The calculated values for contact angles of the PVDF membranes from all performed experiments are presented in Table . The standard error for each test was calculated from the measurements in ten recorded frames for each sample.

Table 7- Contact angle measurements on PVDF membrane after fouling/cleaning tests membranes.

Membrane	Contact angle (°)
Wetted	28.1 ± 3.4
MQ/MQ	20.4 ± 0.9
MQ/NaOCl	16.9 ± 0.5
MQ/2*NaOCl	16.6 ± 0.6
Foulant/MQ	15.6 ± 0.6
Foulant/NaOCl	12.8 ± 0.4

The membrane contact angle was observed to slightly decrease when the membrane was exposed to both foulant and NaOCl. It was expected that the membrane hydrophobicity could increase (i.e. higher contact angle) when NaOCl attack on the membrane surface was expected to remove the hydrophilic surface additives from the membrane, resulting in a more hydrophobic membrane. Moreover, the membrane was fouled due to the deposition of the foulant mixture (humic acid, alginate and BSA). Membrane fouling was reported to result in a more hydrophobic membrane [11].

The results from the Foulant/NaOCl experiment indicated that membrane hydrophobicity was further reduced. However, the membrane contact angle results also depend on the membrane morphological characteristics such as membrane roughness and consequently pore size. It was reported that membranes with larger pores size feature a rougher surface [12]. Alterations in membrane pore size can be responsible for the changes observed in the membrane contact angle. On the other hand, the degradation of the foulants by NaOCl can result in formation of hydrophilic products. The oxidation of organics by chlorine was reported to result in formation of triha-lomethanes (THMs), haloacetic acids (HAAs) and adsorbable organic halogen (AOX). Indeed, the impact of chlorine on natural organic matters has been reported to significantly increase the hydrophilicity of the products resulting from the degradation [13].

3.3.3 TGA

Figure 7 shows the rate of mass loss as a function of temperature for PVDF membranes after fouling/cleaning tests. It was observed that the MQ/MQ and Foulant/MQ membranes exhibited similar degradation profiles. However, the Foulant/MQ sample consists of an additional peak at 500 °C, which most probably resulted from the deposited fouling material on the membrane surface. The reproducibility of the peaks obtained for foulant/MQ sample by TGA analysis was confirmed through the triplicate trails carried out. Decomposition occurring between 388-460°C, with an equivalent 24% mass loss of the MQ/MQ sample, characterises additives such as PVP, which are commonly used by the membrane manufactures to increase the membrane hydrophilicity [4].

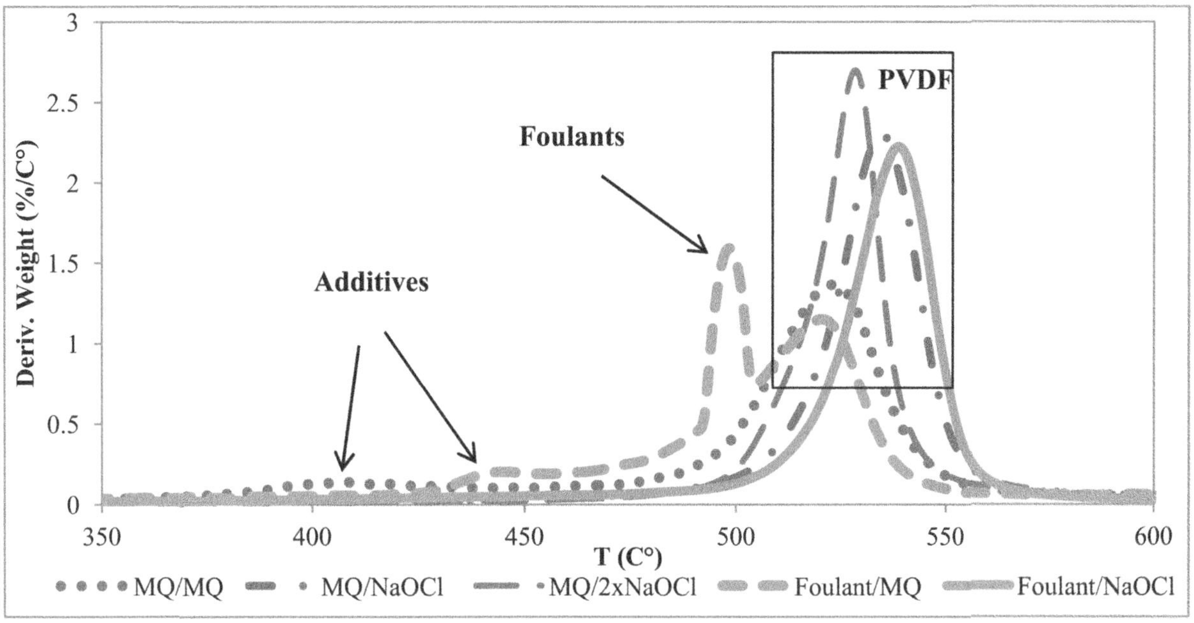

Figure 7- Thermal degradation of fouling/cleaning tests samples.

The mass loss rate curves for all samples challenged by NaOCl were mainly comprised of the degradation step between 500 to 560 °C for PVDF polymer. Therefore, it can be concluded that PVP was mostly removed from any membranes which were in contact with NaOCl. The temperatures of maximum weight loss (T_{max}) in the second degradation step for fouling/cleaning tests samples are summarised in Table 6. The T_{max} peak slightly increased where the PVDF membrane was exposed to NaOCl. The ageing effects of polymers are essentially linked to the modifications of the macromolecular skeleton either through cross-linking or chain scission of the PVDF samples which decreases the impact strength, which can lead to membrane embrittlement [14-15]. The shift of the weight loss peak to higher temperature indicated that the PVDF chains cross link, while chain scission normally leads to a less stable polymeric material, which can degrade at lower temperature [16-18]. However, the T_{max} slightly decreased at the higher cleaning intensity imposed in MQ/2×NaOCl experiment. This can be due to occurrence of chain scission when the membrane was exposed to a higher dose (ppm.h) of NaOCl.

Although cross linking was clearly observed from the results here, a previous study has reported the possible chain scission of PVDF during accelerated ageing with NaOCl. Hypochlorite is a non-specific oxidising agent, which is capable of attacking ether bonds and hydroxyl groups as well as cleaving C-H bonds [18]. Therefore it can be proposed that the ageing process of PVDF here started with cross linking.

Table 6- T_{max} of fouling/cleaning tests membranes in the PVDF region.

Membrane	T_{max} (°C)
MQ/MQ	520
MQ/NaOCl	533
MQ/2×NaOCl	527
Foulant/MQ	518
Foulant/NaOCl	537

3.3.4 FTIR

FTIR analysis of the blank and aged samples was performed and the alterations in membrane function groups were analysed. Figures 8 shows the obtained FTIR spectrums of samples from the fouling/cleaning experiments. Peak suppression was observed between 1660 to 1740 cm^{-1}, which represented the destruction of carbonyl group present in both the PVDF and PVP compounds.

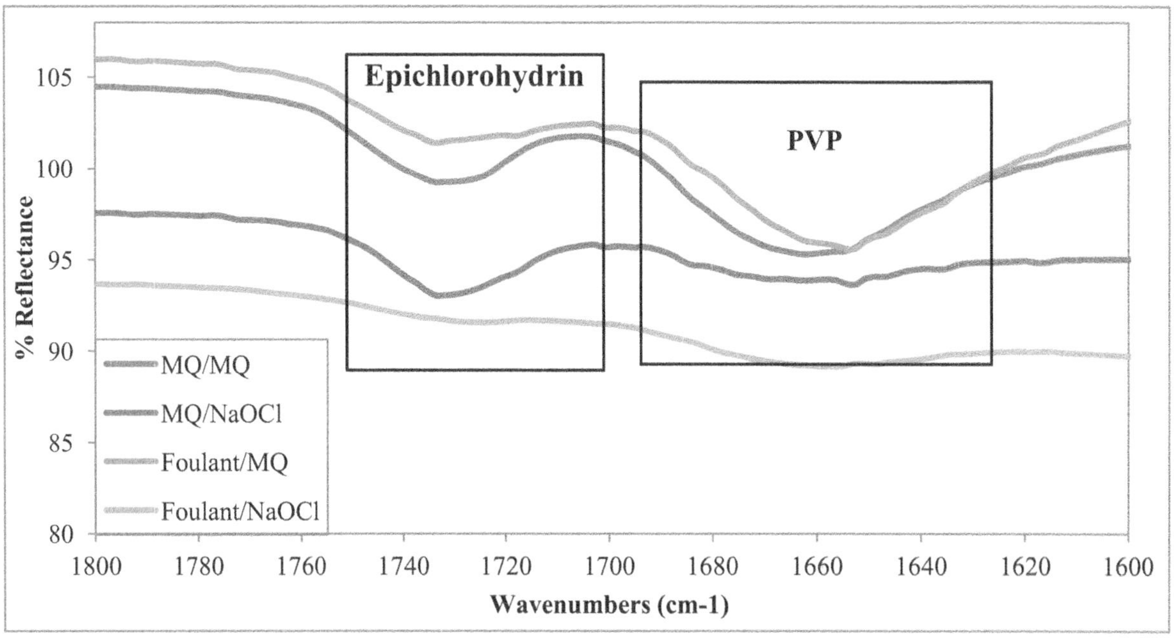

Figure 8- FTIR comparison between blank and Foulant/NaOCl samples.

Table 7- Summary of the alterations in the carbonyl peaks in blank and Foulant/NaOCl samples (✓: similar signal to those obtained with virgin membrane).

Membrane	**PVP**	**Epichlorohydrin**
MQ/MQ	✓	✓
MQ/NaOCl	Suppressed	✓
MQ/2×NaOCl	Suppressed	Suppressed
Foulant/MQ	✓	✓
Foulant/NaOCl	Suppressed	Suppressed

During the Foulant/MQ experiment, no alteration of the PVP was observed due to the presence of foulant. On the other hand, further suppression of both carbonyl peaks were observed for Foulant/NaOCl sample. Although the FTIR spectra support the TGA results regarding the elimination of additives due to NaOCl exposure, the occurrence of cross linking could not be confirmed. Generally, cross linking between the polymer chain results in additional carbonyl bonds, peak growing or broadening of the carbonyl region [19], which could not be clearly observed here. This could be due to overlapping peaks in the FTIR spectrum.

3.3.5 Other tests

Further characterisation techniques such as bubble point and tensile tests were used in this section. The bubble point pressure test can be considered as an indication of the membrane pore size. From Figure 9, the significant decrease in bubble point pressure (from 230 down to around 100 kPa) indicates an apparent increase in the PVDF membrane pore size, when the membrane was exposed to hypochlorite solution. It was also observed that the bubble point pressures of the samples correlated well with the R_{final} values obtained after filtration, both tests revealing a potential pore size enhancement from NaOCl exposure. Additionally, results from contact angle, FTIR and TGA indicated that other physicochemical characteristics of the membrane were altered during the ageing process; potentially impacting the results obtained with the bubble point and resistance tests.

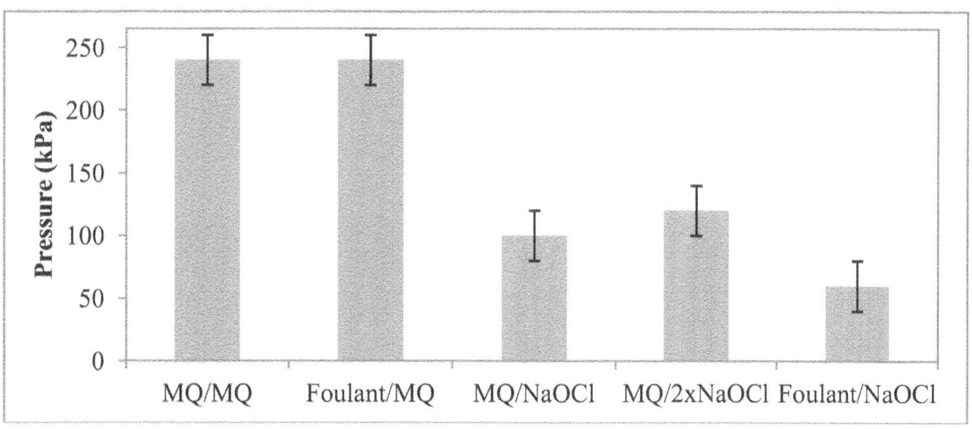

Figure 9- Bubble point pressure of fouling/cleaning tests samples.

The tensile properties of the PVDF membranes were observed to be more severely compromised when the membrane was in contact with hypochlorite only (Table 8). The blank samples, which were exposed to NaOCl, revealed a significant decrease in breaking force and maximum elongation particularly. However, the mechanical properties of the Foulant/NaOCl membrane were not significantly compromised. It was expected that the fouling layer formed under those operating conditions counter-balanced the relative impact of NaOCl by providing mechanical stability to the cleaned membrane. As mentioned in the previous section, FTIR analysis failed to confirm the occurrence of embrittlement due to hypochlorite attack to the membrane polymer. On the other hand, tensile properties of the membrane have been directly linked to membrane ageing in the literature[20]. The embrittlement mechanisms in polymer ageing could be characterised by different behaviours in the stress and yield stress patterns. These reductions signalled a decrease in the elasticity of the membrane. Exposure to NaOCl could result in rearrangement of the molecular structures in the polymeric chains [21]. The tensile results confirmed the increase in degradation temperature observed from the TGA analysis.

Table 8- Tensile properties of PVDF membranes used in benchmark tests and fouling/cleaning cycles.

Membrane	**Breaking force (N)**	**Maximum elong. (%)**
MQ/MQ	2.84	250
MQ/NaOCl	2.76	209
MQ/2×NaOCl	2.02	202
Foulant/MQ	2.96	243
Foulant/NaOCl	2.91	238

3.4 PP membrane ageing

3.4.1 TMP behaviour

The TMP temporal changes of PP membrane against filtration time using feed and 510,000 ppm.h of NaOCl was assessed during consecutive fouling/cleaning cycles (Figure 10). In Figure 10, the solid black line represents the expected TMP behaviour during MQ/MQ tests, which was based on the observed results for PVDF and the R_m values of PP membranes. During the Foulant/NaOCl experiment, the TMP slightly increased to a stable average of 22 kPa, similarly to the results obtained for PVDF membrane in Section 3.3.1.

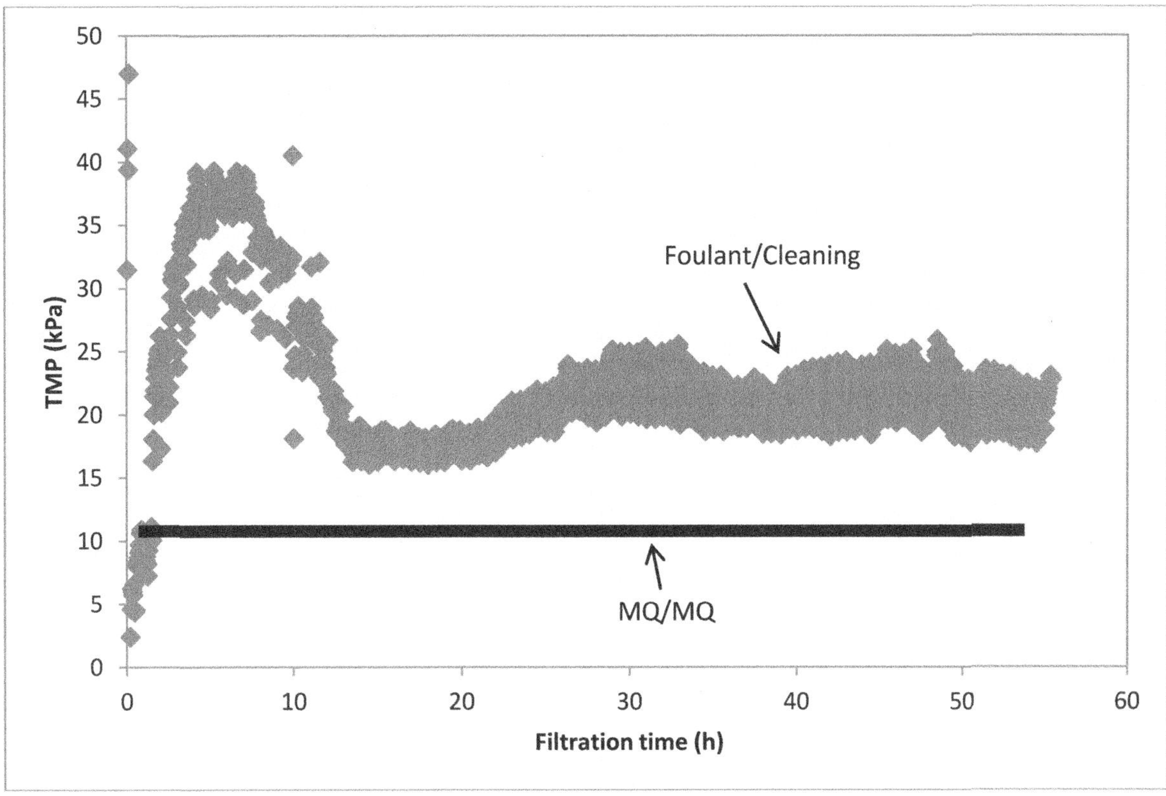

Figure 10- Temporal changes of PP membrane during fouling/cleaning using NaOCl 2%.

The recorded TMP values during the experiment were used to determine the initial and overall fouling rate of PP membrane from Equations 1-2. At the end of the experiment, the membrane was characterised to determine bubble point pressure and TOC rejection level. The characteristics of PP membrane after the fouling/cleaning cycles are summarised in Table 9.

Table 9- Characteristics of PP membrane sample after the fouling/cleaning (NaOCl 2%) cycles.

Membrane	Bubble point pressure (kPa)	dTMP/dt (kPa/h)	$dTMP/dt_{overall}$ (kPa/h)	Rejection (%)
Wetted	200	-	-	22
Foulant/NaOCl	20	9.8	0.277	51

The bubble point pressure of the PP membrane significantly decreased from 200 to 20 kPa, which indicated pore size enlargement, due to NaOCl attack. On the other hand, the rejection of PP increased from 22 to 51% at the end of the test. The higher rejection performance did not correlate with the decrease in bubble point pressure. While the core material of the membrane was expected to be affected by the cleaning, the remaining fouling layer on the surface could be responsible for the increase the apparent rejection. Otherwise, this trend in rejection value did not refute the apparent pore enlargement detected by the bubble point test.

The overall fouling rate of PP membrane was higher than the one for PVDF membrane, indicating that PP was more prone to fouling. On the other hand, the impact of NaOCl attack to PP membrane was more sever comparing to PVDF, as its bubble point pressure decreased more significantly. However, PP membrane pore size was considerably larger (0.2 μm) than those of the PVDF (0.04 μm), which provided more opportunity for fouling material to deposit within the membrane pores and participate actively in rejection.

3.4.2 TGA

Following the filtration experiment, the wetted and Foulant/NaOCl PP membranes were further analysed using TGA. Figure 11 demonstrates the mass loss rate of the PP membrane samples against temperature. PP membrane was represented by only one peak at around 540 °C. If additives and/or preservatives were added during the preparation of these membranes, they were probably degraded around the same temperature range. The PP membrane exposed to NaOCl was observed to degrade at a significantly lower temperature. The maximum weight loss rate temperature of the PP sample decreased from 543 to 488 °C, after the membrane was used in the fouling/cleaning experiment, indicating the reduced thermal stability of the membrane. Degradation at the lower temperature of the same material is a possible indication that the molecular weight of the polymer decreased. In the case of fouled PVDF membrane, a specific degradation peak was detected around 490 °C due to the presence of foulants. Similarly with the PP membrane, it was expected that the peak observed at 488 °C was partly due to the remaining fouling material on the membrane surface.

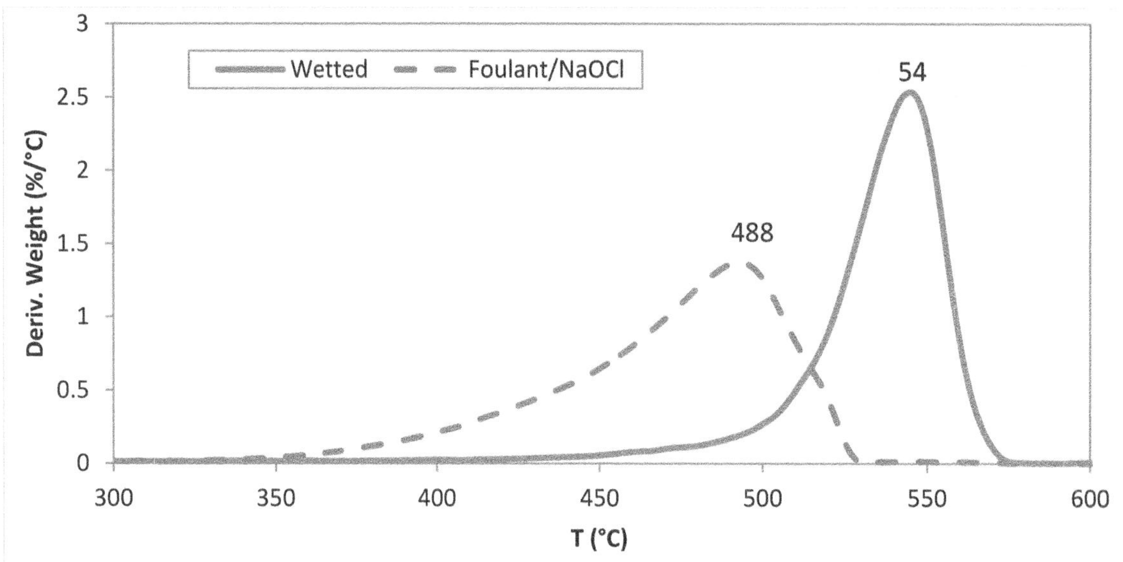

Figure 11-Thermal properties of PP membranes before and after fouling/cleaning cycles.

3.4.3 FTIR

FTIR analyses of wetted and Foulant/MQ samples are shown in Figure 12, where the spectra of the PP membrane after the fouling/cleaning cycles was observed to be altered significantly. A strong peak at 3693 cm^{-1}, representing hydroxyl band was observed on the aged membrane sample. A significant suppression was observed for carbonyl peak at 1741 cm^{-1}. The other carbonyl peak at 1658 cm^{-1} shift to lower wave number (1608 cm^{-1}) and also a new peak emerged at 1510 cm^{-1} in the aged PP FTIR spectrum.

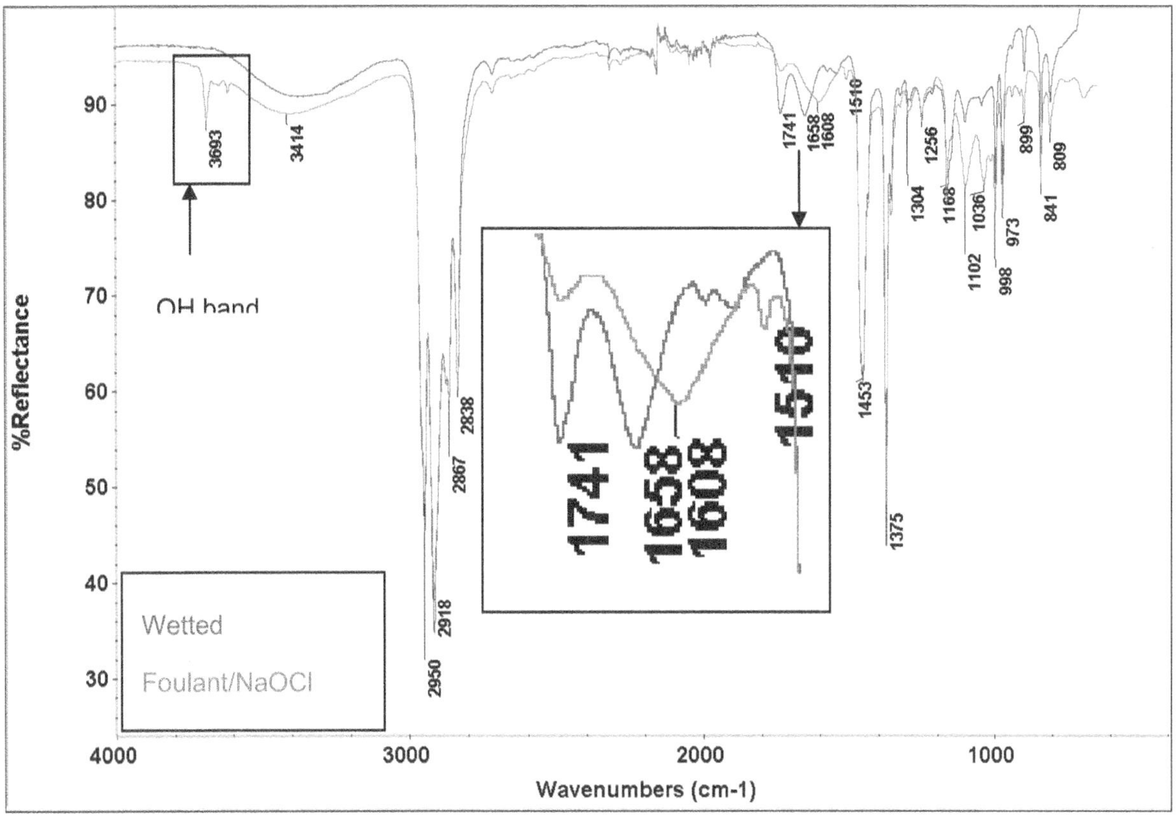

Figure 12- FTIR spectrum of wetted and Foulant/NaOCl PP membranes.

It was previously reported that PP polymer the carboxylic groups converted to carboxylates as the result of oxidative degradation [22]. This conversion was observed to produce a new infrared absorption. Moreover, the new carboxylate absorption band corresponding to the carboxylate salt has been reported to appear between 1550 and 1610 cm^{-1} [22].

The observed hydroxyl band corresponds to carbocylic acid group and, the shifted carbonyl peak (from 1653 to 1608 cm^{-1}) to carbocylate band which confirmed the reported degradation process of PP. Therefore, FTIR analysis of Foulant/NaOCl sample indicated that oxidative degradation and chain scission occurred due to PP membrane exposure to NaOCl. The proposed ageing mechanism was also consistent with TGA results.

3.4.4 Tensile test

The decrease in the mechanical properties of the fouled PP membranes after exposure to NaOCl was showed in Table 10. The breaking force and maximum elongation was reduced more than 80 and 50% respectively, after the fouling/cleaning test. The PP membrane embrittlement could be related to the reduction in molecular weight, affected crystallinity or weaker intermolecular forces in the polymeric chains [23].

Table 10- Tensile properties of wetted and fouling/cleaning cycles of PP membrane samples.

Membrane	Breaking force (N)	Maximum elong. (%)
Wetted	4.95	314
Foulant/NaOCl	1.00	115

4. CONCLUSIONS

A common oxidizing cleaning agent (sodium hypochlorite) was used during consecutive fouling and cleaning cyclical experiments. Following these experiments a range of selected analytical techniques were carried out, allowing a complete assessment of the membrane state after ageing. Benchmark testings were also carried out in order to (1) identify the most relevant membrane property capable to indicate nascent changes in structural changes and (2) assess the relative impact of fouling and cleaning on the filtration behaviour.

Under the operating conditions applied in this study, it was observed that NaOCl recovered most of the membrane performances, after fouling. Within the cleaning intensity used in this series of experiments (up to 510,000 ppm.h), the impact of NaOCl on the structure of the PVDF membrane was detected by most of the analytical tools used. Although the intensity of the chemical attack was generally lower than in our previous experiments (see previous report and [6]), mechanisms of early degradation could still be defined from the current results. The apparent pore size enhancement of the PVDF membrane, concluded from resistance and TOC rejection measurements, was due to the combined effect of increase wetting and hydrophilicity of the membrane. The removal of most of the preservative compounds was also probably responsible for this phenomenon. However, FTIR results and the observed initial degradation of additives, used to increase the membrane hydraulic performances, do not support this finding. It is therefore proposed that the additives could play a major role in the initial wetting of the membrane, while their presence during consecutive filtration is less crucial to suitable operation.

Further characterisation of the various membrane samples indicated that NaOCl exposure lead to the potential damage of the PVDF membrane through cross linking. It was also observed that the mechanical properties of the membrane were slightly compromised. However, the impact of NaOCl on PVDF tensile properties was less pronounced for the fouled membrane. This could be due to the potential protective role of the fouling material, as NaOCl is used to oxidise foulant rather than the membrane material.

Similar trends were observed when these experiments were repeated with PP hollow fibre membranes. Decline in PP membrane performances were significantly more pronounced, when compared to those obtained with the PVDF material. After filtration cleaning cycles, it was observed that the membrane resistance and bubble point pressure decreased significantly, indicating occurrence of pore enhancement. For PP membrane, the ageing mechanism potentially includes chain scission as a result of oxidisation of the polymer due to NaOCl attack. Formation of new O-H band and shifting and degradation of the carbonyl group in the FTIR spectrum of the aged membrane confirmed the proposed ageing mechanism. Moreover, the tensile properties of the PP membrane were also severely reduced after ageing.

REFERENCES

[1]. Arkhangelsky, E., D. Kuzmenko, and V. Gitis, Impact of chemical cleaning on properties and functioning of polyethersulfone membranes, Journal of Membrane Science. 305 (2007) 176-184.

[2]. Bégoin, L., M. Rabiller-Baudry, B. Chaufer, M.-C. Hautbois, and T. Doneva, Ageing of PES industrial spiral-wound membranes in acid whey ultrafiltration, Desalination. 192 (2006) 25-39.

[3]. DOW. *Tech Manual Excerpt, FILMTEC Membranes, Handling and Preservation.* Available from: http://www.filmtec.com.

[4]. Peniche, C., D. Zaldívar, M. Pazos, S. Páz, A. Bulay, and J.S. Román, Study of the thermal degradation of poly(<I>N</I>-vinyl-2-pyrrolidone) by thermogravimetry-FTIR, Journal of Applied Polymer Science. 50 (1993) 485-493.

[5]. Mark, J. and Knovel, Polymer data handbook, Oxford University Press New York, 1999.

[6]. Puspitasari, V., A. Granville, P. Le-Clech, and V. Chen, Optimum Cleaning and Ageing Effect of Sodium Hypochlorite on Polyvinylidene Fluoride (PVDF) Membrane, (2010).

[7]. Wang, D., *Hydrophobic membrane having hydrophilic and charged surface and process.* 1992, United States Patent.

[8]. Li, N., C. Xiao, S. An, and X. Hu, Preparation and properties of PVDF/PVA hollow fiber membranes, Desalination. 250 (2010) 530-537.

[9]. Silva, M.F., C.A. da Silva, F.C. Fogo, E.A.G. Pineda, and A.A.W. Hechenleitner, Thermal and FTIR study of polyvinylpyrrolidone/lignin blends, Journal of Thermal Analysis and Calorimetry. 79 (2005) 367-370.

[10]. Miller, F.A. and C.H. Wilkins, Infrared Spectra and Characteristic Frequencies of Inorganic Ions, Analytical Chemistry. 24 (1952) 1253-1294.

[11]. Verliefde, A.R.D., E.R. Cornelissen, S.G.J. Heijman, I. Petrinic, T. Luxbacher, G.L. Amy, B. Van der Bruggen, and J.C. van Dijk, Influence of membrane fouling by (pretreated) surface water on rejection of pharmaceutically active compounds (PhACs) by nanofiltration membranes, Journal of Membrane Science. 330 (2009) 90-103.

[12]. Richard Bowen, W. and T.A. Doneva, Atomic force microscopy characterisation of ultrafiltration membranes: correspondence between surface pore dimensions and molecular weight cut-off, Surface and Interface Analysis. 29 (2000) 544-547.

[13]. Huang, W.-J. and H.-H. Yeh, Reaction of chlorine with NOM adsorbed on powdered activated carbon, Water Research. 33 (1999) 65-72.

[14]. Fayolle, B., L. Audouin, and J. Verdu, Oxidation induced embrittlement in polypropylene -- a tensile testing study, Polymer Degradation and Stability. 70 (2000) 333-340.

[15]. Gryta, M., J. Grzechulska-Damszel, A. Markowska, and K. Karakulski, The influence of polypropylene degradation on the membrane wettability during membrane distillation, Journal of Membrane Science. 326 (2009) 493-502.

[16]. Jung, Y.-J., Y. Kiso, T. Yamada, T. Shibata, and T.-G. Lee, Chemical cleaning of reverse osmosis membranes used for treating wastewater from a rolling mill process, Desalination. 190 (2006) 181-188.

[17]. Fukatsu, K. and S. Kokot, Degradation of poly(ethylene oxide) by electro-generated active species in aqueous halide medium, Polymer Degradation and Stability. 72 (2001) 353-359.

[18]. Wienk, I.M., E.E.B. Meuleman, Z. Borneman, T.v.d. Boomgaard, and C.A. Smolders, Chemical treatment of membranes of a polymer blend: Mechanism of the reaction of hypochlorite with poly(vinyl pyrrolidone), Journal of Polymer Science Part A: Polymer Chemistry. 33 (1995) 49-54.

[19]. Lazzari, M. and O. Chiantore, Thermal-ageing of paraloid acrylic protective polymers, Polymer. 41 (2000) 6447-6455.

[20]. Arkhangelsky, E., D. Kuzmenko, N.V. Gitis, M. Vinogradov, S. Kuiry, and V. Gitis, Hypochlorite Cleaning Causes Degradation of Polymer Membranes Tribology Letters. 28 (2007) 109-116.

[21]. Gaudichet-Maurin, E. and F. Thominette, Ageing of polysulfone ultrafiltration membranes in contact with bleach solutions, Journal of Membrane Science. 282 (2006) 198-204.

[22]. Scheirs, J., Compositional and failure analysis of polymers: a practical approach, John Wiley & Sons Inc, 2000.

[23]. New Logic Research. *Membrane Cleaning 101*. 2007 [cited 2009 6th August]; Available from: http://membranecleaner.com.

Appendix 4

Fouling/Cleaning Consecutive Testing on Industrially-Aged Membranes
Report No 5

Pierre LeClech

UNESCO Centre for Membrane Science and Technology, University of New South Wales

EXECUTIVE SUMMARY

In this study, a series of consecutive fouling/cleaning cyclical experiments tests were conducted to characterise industrially-aged polypropylene (PP) membranes. The results obtained for the new PP membrane in previous work package were then benchmarked with the behaviour of industrially-aged membranes to better characterise the state of degradation experienced by these membranes. A range of conventional analytical techniques were also used to further characterise the nature of the new, accelerated and industrially-aged membranes.

Compared to the new PP membranes, the industrially-aged membranes presented a lower operating transmembrane pressure (TMP), which can be due to the potential modification of the membranes properties during their lifetime. From these results, it was observed that the industrially-aged membranes featured similar hydraulic performances, which were not significantly affected by the accelerated fouling/cleaning testing. The final total organic carbon (TOC) rejection of all aged membranes improved as the remaining cake layer on these membranes acted as an additional barrier. The formation of different types of fouling on the tested membranes was expected to be responsible for the observed variations in the TOC rejection performances obtained between these membranes.

Further characterisation of membrane samples before and after NaOCl exposure during the fouling/cleaning consecutive testing indicated the potential damage of the PP membrane. Thermogravimetric analysis (TGA) indicated that the received industrially-aged membranes were degraded, possibly through chain session during their long-term use. The extent of the apparent degradation varied between the two membranes. The initial assessment of the industrially-aged membranes via fourier transform infrared spectroscopy (FTIR) indicated that, even though the membrane additives still remained after the long-term industrial use different extents of membranes degradation was evident: the formation of new O-H band, along with shifting and degradation of the carbonyl group (as observed during FTIR analysis of the aged membrane) confirmed oxidation of the membrane material.

The consecutive cyclical ageing test resulted in further instability in membrane nature and thus, decreased molecular weight of membrane material through oxidative degradation due to NaOCl attack. This confirmed that the effect of cleaning agent nature and concentration were significantly greater than the influence of ageing time on membrane degradation. The observed embrittlement of these membranes confirmed the occurrence of the oxidative degradation and chain scission. Indications of pore enhancement due to NaOCl exposure were also clearly observed.

The comparison of the results from the new and industrially-aged membranes indicated that a number of limitations can be addressed in terms of prediction of the large-scale membrane ageing using accelerated ageing trends. For instance, since the cleaning chemicals that were applied in the industry were different to the ones used in this study, results can only be compared to a limited extend. Moreover, the limitations in the maximum concentration and time to be applied during accelerated ageing studies can create obstacles to match the cleaning intensities that are generally used in large-scale plants. Finally, the lack of sufficient information regarding the detailed cleaning protocols used in the industry indeed limits the evaluation of membrane ageing for industrially-aged membranes. Although, the results from these studies do not always reflect the long-term cleaning applied in large-scale plants, they remain useful in terms of investigating degradation mechanisms.

CONTENTS

FIGURES

TABLES

ABBREVIATIONS

AFM	- atomic force microscopy
ATR-FTIR	- attenuated total reflection - Fourier Transform infrared spectroscopy
CA	- cellulose acetate
CIL	- cleaning in line
CIP	- cleaning in place
CLSM	- confocal laser scanning microscopy
EDTA	- ethylene diamine tetra acetic acid
FS	- flat sheet
GPC	- Gel permeation chromatography
HF	- hollow fiber
IC	- inorganic carbon
LMH	- Lm-2h-1
MBR	- membrane bioreactors
MF	- microfiltration
NaOCl	- sodium hypochlorite
PAN	- polyacrylontrile
PE	- polyethylene
PEG	- poly ethylene glycol
PES	- polyethersulfone
PP	- polypropylene
PS	- polysulfone
PVDF	- polyvinylidene fluoride
PVP	- polyvinylpyrrolidone
Rm	- membrane hydraulic resistance
SDS	- sodium dodecyl sulphate
SEM	- scanning electron microscopy
TC	- total carbon
TGA	- thermogravimetric analysis
TMP	- trans membrane pressure
TOC	- total organic carbon
UF	- ultrafiltration
WT	- water treatment
WWT	- wastewater treatment
XPS	- X-Ray photoelectron spectroscopy

1. INTRODUCTION

One of the difficulties faced by operators during membrane autopsy is to assess the exact extent of the damage encountered by the membrane during its lifetime. During the analysis of old membranes, only a couple of parameters are usually measured, creating difficulties in determining the physical and chemical changes which occurred in the nature of the aged membrane. By analysing the detailed characteristics of various membranes during accelerated ageing, it is expected that enough relevant information could be available to estimate the exact advancement of membrane ageing of an old membrane and to use those tests as benchmark for ageing assessment.

In this report, comparison between the behaviour of industrially-aged membranes and results obtained in previous work package have been presented to better characterise the state of degradation experienced by these membranes. In addition to the series of conventional tests used to characterise industrially-aged membranes, the fouling/cleaning consecutive testing was also conducted on those old membranes to assess their potential to sustain further ageing in a controlled (but accelerated) environment.

A common oxidizing cleaning agent (sodium hypochlorite, NaOCl) was used during consecutive fouling/cleaning cyclical runs. By using a range of carefully selected analytical techniques, the complete assessment of the membrane nature was carried out before and after ageing. It was expected that better understanding of their state of ageing could be obtained from the analysis of their further ageing within a well controlled environment.

2. MATERIALS AND METHODS

2.1 Membranes

The study was based on new micro-porous hollow fibre polypropylene (PP) and two industrially-aged membranes (A and B). All three membranes were supplied by Siemens (Memcor) and featured pore diameter of 0.2 µm, outer and inner diameters of 0.6 and 0.3 mm. Table 1 shows the abbreviations used in this study to characterise the state of the various membranes.

Table 1- List of PP membranes and their identifications used in this study (before and after consecutive fouling/cleaning cyclical testing).

Membranes	Before testing	After testing
New	PP_{inital}	PP_{aged}
From plant A	$A_{initial}$	A_{aged}
From plant B	$B_{initial}$	B_{aged}

Membrane A was received from a 7.5 MLD (10^6 L per day) water reclamation plant, after 10 years of operation, where flux recovery was obtained through back washing and chemical cleaning. Chemical cleaning was performed fortnightly using EX-A3 chemical (caustic 50% plus wetting agents) [1]. The cleaning intensity experienced by membrane A was not estimated due to lack of information of the chemical cleaning duration applied in this particular large-scale plant.

Membrane B was received from a 9 year old microfiltration water treatment plant which undergoes regular cleaning at intervals of 250 h operating time. The cleaning protocol consisted of acid cleaning, followed by basic cleaning. The acidic clean consisted of 0.05% sulphuric acid with 0.1% EDTA at 38°C for 6 h whilst the basic clean used a 1% caustic solution with either 1% hydrogen peroxide or 0.5% MemClean (proprietary cleaning solution from Siemens) C at 38°C for 10 h.

The estimated cleaning intensity experienced by membrane B included approximately 296 cleaning cycles which correspond to the total cleaning time of 1,778 and 2,964 h of sulphuric acid and caustic cleaning, respectively. Therefore, it was estimated that membrane B experienced 8.89×10^5 and 2.96×10^7 ppm.h of acidic and caustic cleaning during the 9 years of operation, respectively.

2.2 Module preparation

Tubular cross-flow membrane modules provided the opportunity to use fresh cleaning solution, and also maintain the feed solution concentration relatively constant during the experiments. These membrane modules were composed of 6 parts:

- Two short PVC tubes (±6 cm) with 12.8 mm outside diameter (OD),
- Two union tees connecters,
- One long (9.6 mm ID, 12.8 mm OD) transparent pipe (length depends on demand),
- Hollow fibre membrane, which length should be 3-4 cm longer than the total length of module for potting purpose.

The fibres were inserted inside the transparent tube, where both sides of the fibres were glued using Araldite inside the short pipes. The tubes were then cut at the position shown below and the end of fibres at one side was blocked using glue. The module was then cut T the black line to produce adequate permeate channels (Figure 1).

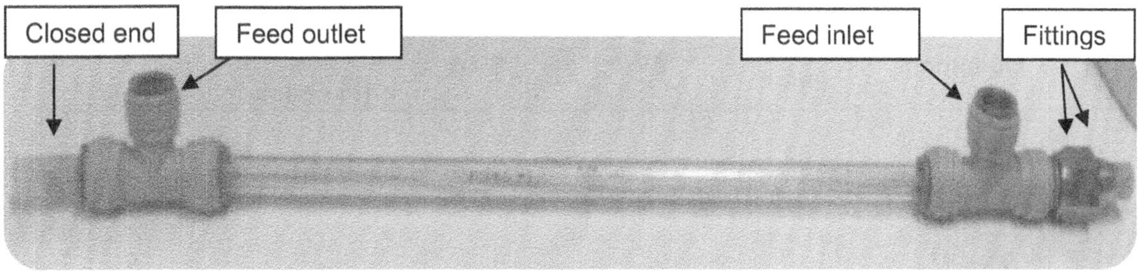

Figure 1- Picture of the hollow fibre module used in the cycling experiments.

2.3 Chemicals

Sodium hypochlorite (NaOCl) from Ajax Finechem with nominal concentration of 12.5% w/v was used as the chemical cleaning agent in the cyclical runs. Humic acid sodium salt, alginic acid sodium salt and bovine albumin serum (BSA) were purchased from Sigma Aldrich. Humic acid, alginate and BSA were used to prepare the feed solution at concentrations of 20, 10 and 10 mg/L respectively.

2.4 Filtration set up and automatic operational phases

In this series of experiments, the membrane fouling/cleaning operation was conducted using the filtration set up shown in Figure 2. The setup was composed of two pumps, two 3-way solenoid valves (SV), one pressure gauge, one pressure transducer, one membrane module and four solution vessels.

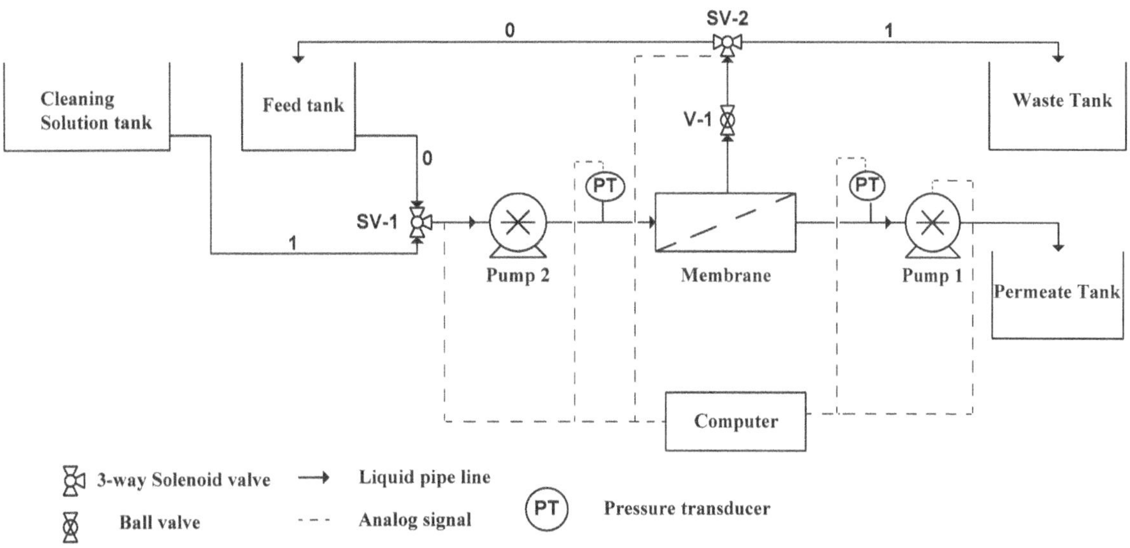

Figure 2-. Schematic illustration of the cyclical fouling/cleaning setup.

All the equipment was pc-controlled. Table 2 shows the state of the pump and the opening position of the two valves (0 or 1, see Figure 1 for stream identification) during the filtration (#1), cleaning (#2), assisted cleaning by filtration (#3) and rinsing (#4) phases. Due to time restrain, the chemical cleaning time was artificially long and extensive, compared to the duration of the filtration (Table 1). A relatively long rinsing period was also necessary to remove all trace of NaOCl before the filtration was resumed. During rinsing time, Pump 2 was set to a higher speed (60 L m^{-2} hr^{-1}) to insure the complete removal of cleaning solution from the module and prevent its recirculation back into the feed tank. The feed stream flux was adjusted manually and was maintained at constant rate during the whole operation. Feed solution was used to rinse the cleaning solution from the module after cleaning phase. Permeate flux rate was fixed at 50 L m^{-2} hr^{-1}.

Table 2.- Operational phases and position of the solenoid valves during each fouling/cleaning cycle.

#	Operation	Duration (min)	SV-1	SV-2	Pump 1 (Permeate)	Pump 2 (Feed)
1	Filtration	30	0	0	on	on
2	Cleaning	17	1	1	off	on
3	Assisted cleaning	2	1	1	off	on
4	Rinsing	10	0	1	off	on

2.5 Membrane characterisation

Following the ageing process, different membrane characterisation techniques were adopted to investigate the suitability of the each method in assessing membrane degradation. In the previous study (work package #2), the suitability and reliability of the different membrane characterisation techniques in detecting membrane degradation were assessed and compared. The comparison was performed by developing a ranking system, with each criterion being given a weighting factor depending on its importance.

Table 3 shows the calculated reliability of the selected membrane characterisation techniques. These tests were carried out on five new membrane samples. The average value of the samples was calculated for each test. The sample standard deviations of each characterisation method were then computed to find the variability in results from the mean value.

Table 3- Summary of the reliability assessment of analytical techniques used on new membrane.

Characterisation techniques	Standard deviation (%)
Clean water test	6
Bubble point test	20
Maximum elongation	10
TGA	1

2.5.1 Membrane resistance and TMP behaviour

Clean water test was performed after wetting and rinsing the membrane to measure its intrinsic resistance (R_m). The clean water test for the hollow fibre membranes in the tubular module was performed using the flux stepping method. Transmembrane pressure (TMP) was calculated as the difference in pressure between the feed and the permeate side of the membrane measured by pressure transducers. Constant flux filtration was carried out at 20 $Lm^{-2}h^{-1}$, and increased to 50 and 100 $Lm^{-2}h^{-1}$ after 30-60 s intervals, while the pressure was recorded for each flux step.

Figure 3 illustrates the TMP behaviour of hollow fibre membrane during a typical filtration and cleaning run. During the fouling/cleaning experiments, the TMP behaviour during each filtration cycle was expressed as fouling rate (dTMP/dt) and was calculated using Equation 1:

$$\frac{dTMP}{dt} = \frac{TMP_b - TMP_a}{t_b - t_a} \qquad \text{Equation 1}$$

Where TMP_a and TMP_b are the initial (after 10 min of filtration) and final TMP (kPa) of the filtration cycle, respectively. The t_a and t_b as their corresponding times (h). The overall TMP increase rate during the entire filtration time of these experiments ($dTMP/dt_{overall}$) was calculated using Equation 2.

$$\frac{dTMP}{dt}_{overall} = \frac{Average(TMP_d, TMP_e, TMP_f) - Average(TMP_a, TMP_b, TMP_c)}{t_f - t_a} \qquad \text{Equation 2}$$

Where $TMP_{d\ to\ f}$ and $TMP_{a\ to\ c}$ represent the final TMP of the last and first 3 three filtration cycles.

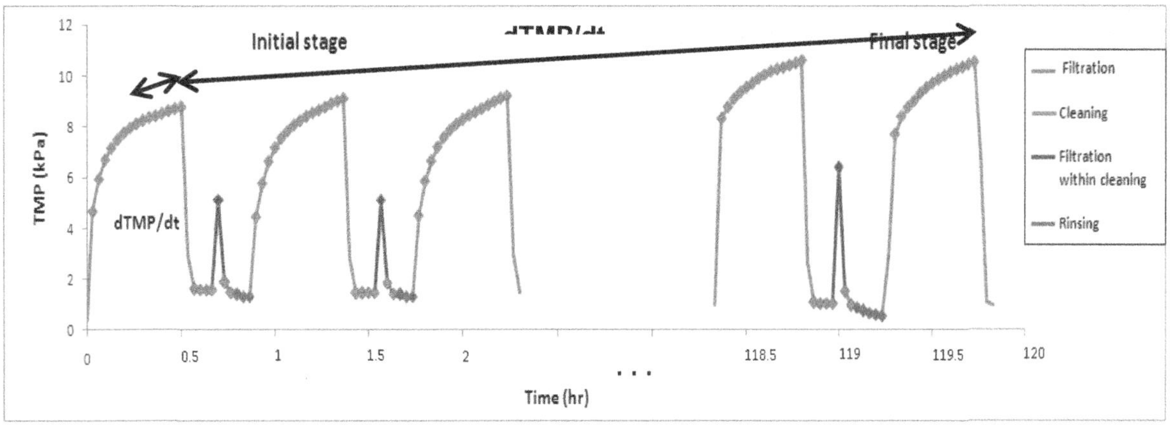

Figure 3- TMP behaviour and fouling rates (dTMP/dt) calculations during initial and final stages during fouling/cleaning cyclical experiment.

2.5.2 Tensile test

Measurements of the mechanical properties of the membranes were performed with tensile tests. The material was clamped at both ends and stretched at constant speed. The force required to elongate the material with time was recorded until the material breaks. The tests were performed on a TA.XT2i/5 texture analyser (Texture Technologies), which has a maximum pulling force of 50 N. Hollow fibre samples were cut into 40 mm length. The test speed and the initial distance were 0.5 mm/s and 35 mm respectively. The software used to collect information from the Texture Analyser is Stable Micro Systems® Texture Expert. It recorded the force needed to elongate the material and the distance between grips per sampling time and plotted the graph of force (N) against time (s).

2.5.3 Thermogravimetric analysis (TGA)

Thermogravimetric analysis (TGA) is a thermal analysis technique which measures the weight change in a material as a function of temperature and time in a controlled atmosphere. The technique is useful when investigating the behaviour and thermal stability of materials, as well as performing compositional analysis of materials. Qualitative and quantitative analysis could be carried out with minimal solid sample size, with measurement range between 0.1 and 200mg. TGA tests in this study were performed on the Thermal Advantage® device manufactured by TA Instruments. The samples were dried prior to testing to prevent the presence of moisture in the samples. Analysis was carried out in inert atmosphere (N_2 gas), with temperature increase in the range of 20-1000°C. Test method chosen was the Ramp method, which increases the temperature 20°C/min.

2.5.4 Fourier transform infrared spectroscopy (FTIR)

FTIR allows the detection of changes in the functional groups of the membrane by comparing membranes before and after exposure to cleaning agents. FTIR detects bonds between atoms in a molecule and can thus provide information about functional groups and chemical structures of the membrane. Each peak illustrates a specific type of molecular vibration: symmetric, asymmetric, twisting, wagging, bending or rocking. The FTIR spectra were analysed with the high resolution Hummel polymer and additives spectrum library available with the FTIR program. Bruker IFS66/S FT/IR Spectrometer was applied to obtain the spectrums for new and aged membranes.

2.5.5 Bubble point test

To determine the bubble point pressure of the hollow fibre membranes, the potted modules were completely filled with Milli-Q water. Then, nitrogen gas was connected to the feed inlet side of the module. The pressure was read from a pressure indicator attached between the gas cylinder and the membrane module. During the test, the pressure was gradually increased until the first bubble was observed through the membrane surface.

3. RESULTS AND DISCUSSIONS

3.1 Membrane ageing

In this section, the results from industrially-aged membranes A and B are presented before and after fouling/cleaning challenge test. The characteristics of new PP membrane, which was also aged through cyclical fouling/cleaning test (work package # 4) is also reported as a benchmark.

3.1.1 TMP Temporal changes during cyclical testing in industrially-aged membranes

The TMP temporal changes of new, A and B membranes obtained during consecutive fouling/cleaning cycles using the foulant mixture and 510,000 ppm.h of NaOCl are summarised in Figure 4. Each data point in the Figure 4 is an average of recorded TMPs during 2 min. The recorded TMP values during these experiments were used to determine the final and overall fouling rate of these PP membranes from Equations 1-2. Table 4 reports the average of the last 3 filtration cycles of the experiments.

The observed trend of the TMP behaviour of the new membrane showed a significant increase to 37 kPa during the initial stages (first 10 h) of the test, which was followed by a sharp decrease to 21 kPa. The rapid initial TMP increase observed for the new membrane can be as the result of poor system stability. Throughout the rest of the operating time, the TMP of the new membrane gradually increased to 23 kPa. The membrane A TMP increased gradually from approximately 11 to 15 kPa through the entire operating time. The TMP increase observed for sample B was relatively lower than the ones observed for new and A membranes. The TMP trend for B showed a slight decrease from 7 to 5 kPa during the initial stages of the experiment, which was followed by a gradual increase to 9 kPa at the end of the operating time.

From these results, it appeared that both membranes A and B featured similar performances and do not seem to be significantly affected by the fouling/cleaning testing. In comparison to the behaviour of brand new PP membranes, the industrially-aged membranes presented a lower operating TMP, which can be due to potential modification of these membranes properties during their lifetime.

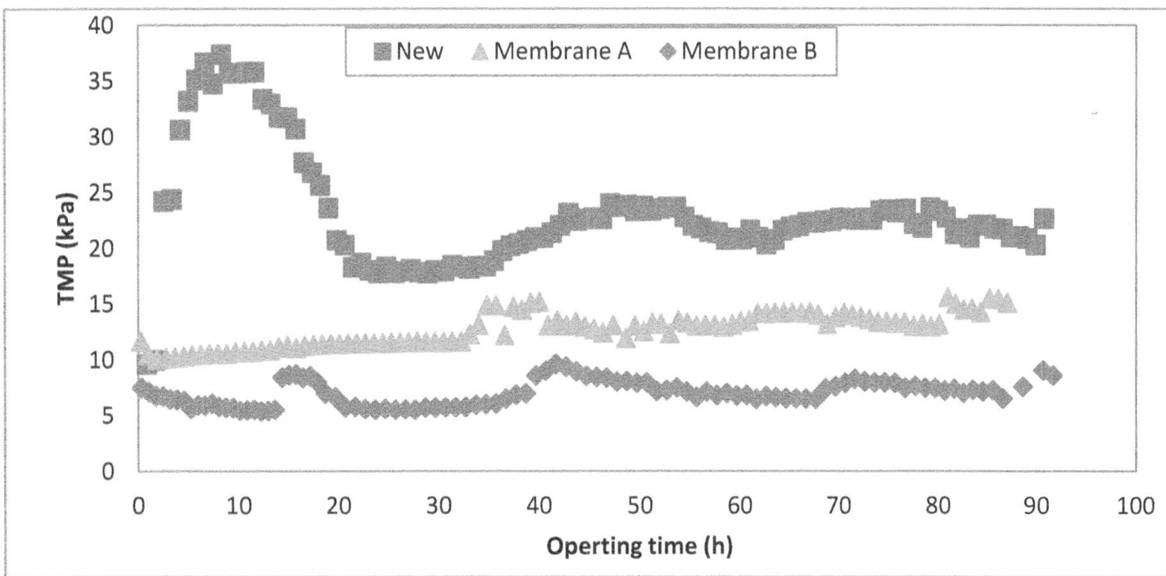

Figure 4- Temporal changes of PP new, A and B membranes during fouling/cleaning using NaOCl 2%.

The new membrane resistance and the dTMP/dt values obtained during the fouling/cleaning test increased significantly over the operating time. The highest resistance and TMP_{final} and $dTMP/dt_{overall,}$ was observed for the new membrane, as it was more severely fouled than the other tested membranes in this study. As can be seen in Table 4, the resistance of membrane A increased from 3.3 to 6.9×10^{11} m^{-1} at the end of the experiment, indicating that even though chemical cleaning was performed, the foulants were not completely removed. As expected, fouling resulted in the increase of the TMP_{final} to 15 kPa and $dTMP/dt_{overall}$ to 0.1 kPa/h. Moreover, membrane A featured the highest fouling rate (12 kPa/h) within the final filtration cycles amongst the tested samples. The membrane B fouling behaviour (Table 4) exhibited similar trends to membrane A nevertheless; the R_{final}, TMP_{final} and the final and overall fouling rates featured smaller values in this case.

Table 4- Hydraulic and rejection performances for new and industrially-aged membranes during fouling/cleaning cyclical experiments.

Membrane	$R_m \times 10^{-11}$ (m^{-1})	$R_{final} \times 10^{-11}$ (m^{-1})	TMP_{final} (kPa)	dTMP/dt (kPa/h)	$dTMP/dt_{overall}$ (kPa/h)	$Rejection_{final}$ (%)
New	6.2	14.4	23	9.8	0.28	51
Membrane A	3.3	6.9	15	12.0	0.10	61
Membrane B	3.5	6.3	9	8.1	0.07	46

The higher overall fouling rate of new PP compared to the industrially-aged membranes, indicated that the new membrane was more prone to fouling. The dTMP/dt values indicated that the dominating fouling mechanisms for these three PP membranes varied. According to the fouling behaviour of these samples, it seems that the new membrane was fouled mainly through internal fouling, whereas the dominant fouling mechanism for membrane A and B was probably cake formation. This conclusion was drawn from previous observations indicating that fouling through pore restriction was more persistent to cleaning, while the cake layer on the membrane surface was found to be partly reversible [2]. Moreover, higher fouling rate of microporous membranes was reported to increased membrane resistance more rapidly in the case of pore closure and cake formation rather than that of pore narrowing [3]. This can explain the higher overall fouling rate and the lower dTMP/dt within the filtration cycles of the new PP compared to membrane A, since internal fouling is generally more difficult to remove, compared to the other types of fouling.

At the end of each experiment, the aged membrane was also characterised in terms of TOC rejection level (Table 4). The initial TOC rejection of all three membranes was assessed in the first filtration cycle of these experiments. Surprisingly, the combined effect of fouling and cleaning during the industrial use of membrane A and B resulted in the same initial rejection performance (21 ± 1%) as the new PP membrane. The new, A and B membranes final TOC rejections significantly increased to 51, 61 and 46% respectively. The increase in the final TOC rejection of all the tested membranes was probably due to the incompelete removal of foulants from the membrane.

The final TOC rejection of new membrane A was lower than the one observed for sample A, as the remaining cake layer on membrane A acted as an additional barrier, resulting in a better final TOC rejection. The porous cake layer formation which participated to the membrane final TOC rejection resulted in only slight increase in membrane final resistance. Although the hydraulic performances of A and B were quite similar, their final rejection levels varied significantly. Formation of different types of fouling on A and B can be responsible for the observed variations in the TOC rejection performances.

3.1.2 TGA

Following the filtration experiments the new, A and B membranes were further analysed using TGA. Figure 5 demonstrates the mass loss rate and the maximum weight loss rate temperature (T_{max}) of the new membrane before and after fouling/cleaning cycles ($PP_{initial}$ and PP_{aged} respectively), as well as samples $A_{initial}$ and $B_{initial}$. To assess the degradation extent of industrially-aged membrane samples ($A_{initial}$ and $B_{initial}$), the $PP_{initial}$ and PP_{aged} samples were used as a benchmark. The TGA curve of $PP_{initial}$ sample showed only one sharp peak at around 540 °C. If additives and/or preservatives were added during the preparation of these membranes, they were probably degraded around the same temperature range. After the fouling/cleaning test the PP_{aged} membrane T_{max} was observed to significantly shift to a lower temperature. Moreover, the thermal degradation curve of PP_{aged} featured a significantly broader peak, which can be due to impurities in the sample resulting from irreversible fouling using the cyclical testing. Thermal degradation of the membrane at the lower temperature can be an indication of lower molecular weight of the polymer or chain scission. The T_{max} of $A_{initial}$ was observed at a significantly lower temperature (518 °C) than the one for $PP_{initial}$, indicating the reduced thermal stability of the membrane. The T_{max} of membrane $B_{initial}$ sample was observed at a yet lower temperature (508 °C) than $A_{initial}$.

On the other hand, both $A_{initial}$ and $B_{initial}$ peaks presented an intermediate profile than the initial and aged PP T_{max} values and were slightly broader. Therefore, the short-term fouling/cleaning experiment was observed to have a stronger impact on the PP membrane thermal stability than the long-term industrial use experienced by samples $A_{initial}$ and $B_{initial}$. This confirms that the effect of cleaning agent nature and concentration on membrane degradation are significantly greater than the influence of time, which was consistent with the reported observation in the literature [4]. The broader peaks observed for $A_{initial}$ and $B_{initial}$ peaks were probably due to broken PP chains or to the remaining foulants on these membranes after industrial use or [5].

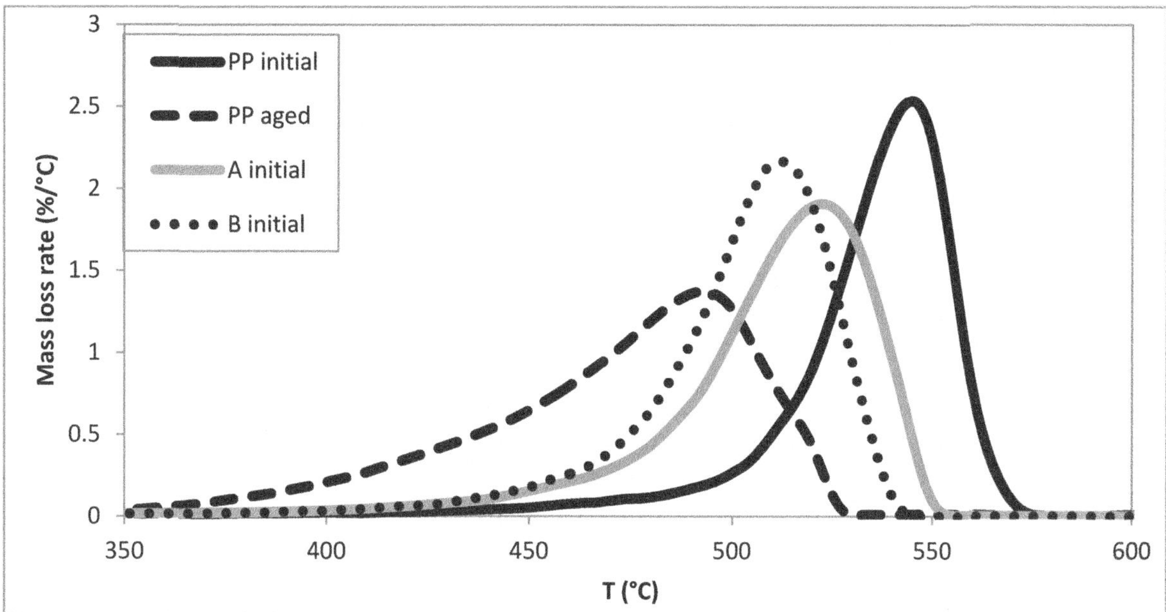

Figure 5- TGA results of $PP_{initial}$, $A_{initial}$, $B_{initial}$ membranes, as well as PP_{aged} after 2% NaOCl cleaning.

Figure 6 shows the mass loss rate of the PP_{aged}, A_{aged} and B_{aged} samples versus temperature. The T_{max} of A_{aged} further decreased to 503 °C, which was still greater than the one for PP_{aged}. Following the cyclical fouling/cleaning, the thermal degradation peak of membrane B sample appeared significantly broader (similar to PP_{aged}) and with a slightly lower T_{max} (493 °C). As mentioned above, broad thermal degradation peak generally suggests the presence of impurities in the sample. These impurities can comprise of both remaining foulants from the cyclical testing and/or broken PP polymer chains. On the other hand, the TGA curve of A_{aged} sample was not further broadened, although the remaining foulants on the membrane were expected to be observed in the TGA results. Therefore, it is more likely that the degradation at the lower temperature (375 °C) of B_{aged} membrane was due to the occurrence of chain scission.

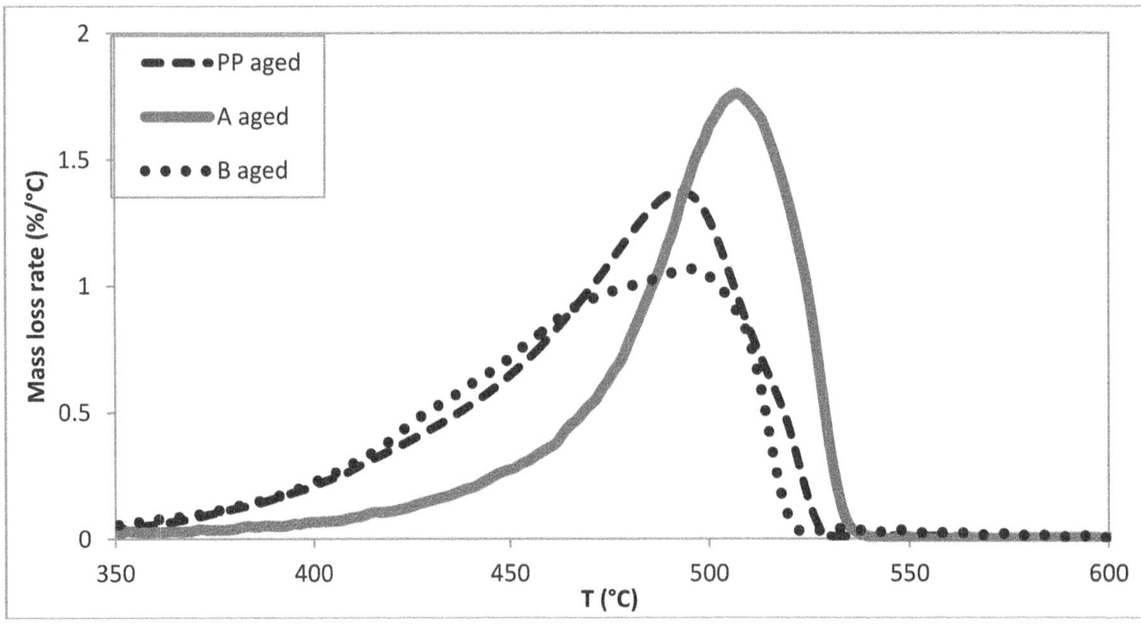

Figure 6- TGA results of PP_{aged}, A_{aged} and B_{aged} membranes after 2% NaOCl cleaning.

3.1.3 FTIR

All the membranes were further analysed using FTIR before and after the cyclical ageing experiments (Table 5). The spectrum of the $PP_{initial}$ membrane showed distinct carbonyl peaks at 1658 and 1738 cm^{-1}, which represent the membrane additives, such as PVP [6]. The spectrum of the $A_{initial}$ membrane also featured both carbonyl peaks similar to $PP_{initial}$, though the peaks at 1658 and 1738 cm^{-1} appeared in higher lower and intensities, respectively. This indicated that even after the long-term industrial use of membrane A, the additives still remained in the membrane material. Furthermore, $A_{initial}$ spectrum showed emerging peaks at 1543 and 3690 cm^{-1}, representing carboxylic acid group and hydroxyl respectively. These peaks possibly resulted from the presence of the membrane degradation products. The $B_{initial}$ spectrum prior to the cyclical tests was similar to $A_{initial}$ sample. Then again, small changes, such as different intensities of the carbonyl peaks and an emerging hydroxyl band, were observed in the $B_{initial}$ spectrum.

Table 5- FTIR peak values and their corresponding functional groups for all the tested membranes before and after cyclical cleaning.

Wave number (m^{-1})	PP_{inital}	PP_{aged}	$A_{initial}$	A_{aged}	$B_{initial}$	B_{aged}
PP Finger print						
700-1100	✓	Alternated	Alternated	Alternated	Alternated	Alternated
Additives						
1658	✓	N.d.	+	N.d.	+	N.d.
1738	✓	-	-	-	-	-
Degradation products						
1543	N.d.	✓	✓	✓	N.d.	✓
3620	N.d.	✓	N.d.	+	✓	+
3690	N.d.	✓	✓	+	✓	+

✓: Peak is present, +: Peak intensity increased, -: Peak intensity decreased, N.d. : Not detected.

The spectra of all tested membranes after the fouling/cleaning cycles were altered significantly. A strong peak at 3690 cm^{-1}, representing hydroxyl band was observed in all aged membrane samples. A significant suppression was observed for carbonyl peak at 1741 cm^{-1} for PP_{aged} membrane. Moreover, the other carbonyl peak at 1658 cm^{-1} in the PP_{aged} spectrum shifted to lower wave number (1543 cm^{-1}), and also a new peak emerged at 3960 cm^{-1}. It was previously reported that in the PP polymer, carboxylic groups converted to carboxylates as the result of oxidative degradation [5]. This conversion was observed to produce a new peak. Moreover, the new carboxylate absorption band corresponding to the carboxylate salt has been reported to appear between 1550 and 1610 cm^{-1} [5]. Therefore, FTIR analysis of PP_{aged} sample indicated that oxidative degradation and chain scission occurred due to membrane exposure to NaOCl. The proposed ageing mechanism was also consistent with TGA results.

The hydroxyl band intensity in the A_{aged} spectrum significantly increased, and the carbonyl peak (from 1658 to 1543 cm^{-1}) shifted to carboxylate band, which confirmed that A_{aged} was further oxidated by NaOCl. The spectra of B_{aged} membrane also showed a new hydroxyl band, as well as suppression and shift in the carbonyl peaks to the carboxylate band wave number. This indicated that the B_{aged} membrane has gone through oxidative degradation and consequently chain scission due to NaOCl attack.

It is noteworthy that both A and B membrane initial and aged spectra were significantly different to the new membrane between 1100 to 700 cm^{-1} frequencies, which represents PP finger print. This indicates that the PP membranes produced more than a decade ago and used in the industry since were obtained from a different manufacturing process (i.e. different polymer molecular weight or additives).

3.1.4 Other tests

In order to further investigate the impact of ageing on the membrane pore size and tensile properties, bubble point pressure (P_{bp}) and maximum elongation (E_{max}) obtained during tensile testing of the samples were determined before and after the cyclical experiments. Table 6 shows the E_{max} and bubble point pressure values of all the membranes in this study.

The P_{bp} of $A_{initial}$ membrane (220 kPa) did not indicate any changes in the sample pore size. On the other hand, after the fouling/cleaning cycles, P_{bp} of the A_{aged} sample decreased to 20 kPa. This indicated that the pore size of the membrane was significantly compromised due to NaOCl exposure.

The observed P_{bp} trend for membrane B sample varied from the one for sample A. For example, the P_{bp} of $B_{initial}$ was considerably lower than the other two membranes. Furthermore, P_{bp} did not seem to change after the ageing experiment. Given that the both A and B membranes were PP and manufactured by Memcor, the variations in the initial P_{bp} of the industry aged membranes was probably due to the different cleaning agents or feed waters applied to these membrane during their large-scale operation. Still, it is surprising to observe such significant variations between $A_{initial}$ and $B_{initial}$ interms of P_{bp} and E_{max}, given that those two samples presented so many similarities from previous testing.

Table 6- Bubble point pressure (P_{bp}) and maximum elongation (E_{max}) of new, membare A and B before and after cycles. The measured standard deviation of P_{bp} and E_{max} were 20 and 10%, respectively.

Membrane	**P_{bp} (kPa)**	**E_{max} (%)**
$PP_{initial}$	200	314
PP_{aged}	20	115
$A_{initial}$	220	324
A_{aged}	20	104
$B_{initial}$	120	231
B_{aged}	110	168

The E_{max} of virgin membrane decreased from 314 to 115 % after the fouling/cleaning experiment. The embrittlement of the PP membrane confirmed the occurrence of the oxidative degradation of the new membrane and chain scission. The E_{max} of industrially-aged membranes however, followed two different trends. Membrane $A_{initial}$ tensile properties was similar to the $PP_{initial}$ (324 %), which suggested that its tensile properties were not altered during the 10 years of operation. After the fouling/cleaning cycles the E_{max} of membrane A_{aged} was decreased to 104 % as the same manner as the new membrane. On the other hand, the E_{max} value of the $B_{initial}$ was 231%, which was significantly lower compared to $PP_{initial}$ and $A_{initial}$. After the fouling/cleaning cycles, the E_{max} value of the B_{aged} sample decreased to 168%, which was higher than the other two aged samples, indicating a possible conditioning of the membrane which protected the membrane against NaOCl attack during cyclical cleaning.

4. CONCLUSIONS

In this study, a series of consecutive fouling/cleaning cyclical experiments tests using a common oxidizing cleaning agent (sodium hypochlorite, NaOCl) was conducted on two different industrially-aged PP membranes. By using a range of carefully selected analytical techniques, the complete assessment of the membrane degradation was carried out before and after accelerated ageing. The results obtained for the new PP membrane in previous work package were then compared with the behaviour of industrially-aged membranes as a benchmark to better characterise the state of degradation experienced by these membranes.

Compared to the new PP membranes, the industrially-aged membranes presented a lower operating TMP, which can be due to the potential modification of these membranes properties during their lifetime. From these results, it was observed that both industrially-aged membranes featured similar hydraulic performances, which were not significantly affected by the accelerated fouling/cleaning testing. The final TOC rejection of all aged membranes improved as the remaining cake layer on these membranes acted as an additional barrier. Formation of different types of fouling on the tested membranes was expected to be responsible for the observed variations in the TOC rejection performances obtained between the membranes.

Further characterisation of membrane samples before and after NaOCl exposure indicated the potential damage of the PP membrane. TGA analysis indicated that the industrially-aged membranes were relatively degraded, possibly through chain session during their long-term industrial use. The extent of the apparent degradation varied between the two samples, as well as the new PP membrane. Therefore, the PP membranes produced more than a decade ago and used in the industry since, were possibly obtained from a different manufacturing process (i.e. different polymer molecular weight or additive compounds). The initial assessment of the industrially-aged membranes via FTIR indicated that, even though the membrane additives still remained after the long-term industrial use of in the membrane material, different extents of membranes degradation was evident: the formation of new O-H band, along with the shifting and the degradation of the carbonyl group (as observed during FTIR analysis of the aged membrane) confirmed oxidation of the membrane material.

The consecutive cyclical ageing resulted in further instability and thus, decreased molecular weight of membrane material through oxidative degradation due to NaOCl attack. This confirmed that the effect of cleaning agent nature and concentration were significantly greater than the influence of ageing time on membrane degradation. The observed embrittlement of these membranes confirmed the occurrence of the oxidative degradation and chain scission. Indications of pore enhancement due to NaOCl exposure were also clearly observed.

A number of limitations can be addressed in terms of prediction of the large-scale membrane ageing using accelerated ageing trends; including:

1. The nature of the cleaning agent was observed to influence not only the mechanism of the membrane degradation, but also the severity of membrane integrity loss (see work package #3). Therefore, since the cleaning chemicals applied in the industry were different to the ones used in this study, results can only be compared to a limited extend.

2. Limitations with the maximum concentration and time that can be applied during accelerated ageing studies can create obstacles to match the cleaning intensities that are generally used in large-scale plants. Most accelerated ageing studies use high concentrations of cleaning agents in relatively short-term lab-scale experiments to mimic long-term effect of membrane cleaning on membranes. On the other hand, the use of high concentrations of these chemicals has an instantaneous effect on some membranes or experimental apparatus. For example, relatively low concentration of NaOH in this study was applied due to instant reaction of PVDF when higher concentrations were used. Given the limited time available to conduct those accelerated ageing experiments, the overall cleaning intensity experienced in the industry can hardly been reached in the labs.

3. The lack of sufficient information regarding the cleaning protocols used in the industry, including the exact nature of the chemicals (mostly proprietary) and the frequency and the duration of chemical cleaning remains major limitation in this study. This indeed limits the evaluation of membrane ageing for the industrially-aged membranes.

Although the results from these studies do not always reflect the long-term cleaning applied in large-scale plants, they remain important information allowing detailed investigation of the degradation mechanisms.

REFERENCES

[1]. Phuntsho, S., A. Listowski, H. Shon, P. Le-Clech, and S. Vigneswaran, Membrane autopsy of a 10 year old hollow fibre membrane from Sydney Olympic Park water reclamation plant, Desalination. (2011).

[2]. Katsoufidou, K., S.G. Yiantsios, and A.J. Karabelas, An experimental study of UF membrane fouling by humic acid and sodium alginate solutions: the effect of backwashing on flux recovery, Desalination. 220 (2008) 214-227.

[3]. Ye, Y., P.L. Clech, V. Chen, and A.G. Fane, Evolution of fouling during crossflow filtration of model EPS solutions, Journal of Membrane Science. 264 (2005) 190-199.

[4]. Antony, A., R. Fudianto, S. Cox, and G. Leslie, Assessing the oxidative degradation of polyamide reverse osmosis membrane--Accelerated ageing with hypochlorite exposure, Journal of Membrane Science. 347 (2010) 159-164.

[5]. Scheirs, J., Compositional and failure analysis of polymers: a practical approach, John Wiley & Sons Inc, 2000.

[6]. Puspitasari, V., A. Granville, P. Le-Clech, and V. Chen, Optimum Cleaning and Ageing Effect of Sodium Hypochlorite on Polyvinylidene Fluoride (PVDF) Membrane, (2010).

[7]. Li, N., C. Xiao, S. An, and X. Hu, Preparation and properties of PVDF/PVA hollow fiber membranes, Desalination. 250 (2010) 530-537.

IWA Publishing's authorised EU representative for General Product Safety Regulations is Diane D'Arras, 15 rue Duret, 75116 Paris, France, e-mail: safety@iwap.co.uk.

Printed and bound by CPI Group (UK) Ltd, Croydon, CR0 4YY
12/05/2026
02108893-0014